Chrono-topologies

Hybrid Spatialities and Multiple Temporalities

Critical Studies

Vol. 32

Rodopi

Amsterdam - New York, NY 2010

Chrono-topologies

Hybrid Spatialities and Multiple Temporalities

Edited and Introduced by

Leslie Kavanaugh

Book Layout: Janine Toussaint

Cover Photograph: Leslie Kavanaugh

Cover Design: Pier Post

The paper on which this book is printed meets the requirements of "ISO 9706:1994, Information and documentation - Paper for documents - Requirements for permanence".

ISBN: 978-90-420-3141-8
E-Book ISBN: 978-90-420-3142-5

Printed in the Netherlands

Table of Contents

1. Introduction 3
Leslie Kavanaugh

2. Minkowski's Space-Time: From Visual Thinking to the Absolute World
Peter Galison 9

3. Materialist Theories of Time
Richard T. W. Arthur 43

4. Corollaries on Space and Time: A Survey of Arabic Sources in Science and Philosophy
Nader El-Bizri 63

5. Agency and Space in Darwin's Concept of Variation
Chunglin Kwa 79

6. The Time of History/The History of Time
Leslie Kavanaugh 91

7. Places Lived in Time
Mary Lynne Ellis 125

8. Intermittences: Merleau-Ponty and Proust on Time and Grief
Patricia Locke 147

9. Lyrical Bodies: Music and the Extension of the Soul
Sander van Maas 159

10. Phased Space
Raviv Ganchrow 179

11. The Evidence of Film and the Presence of the World: Jean-Luc Nancy's Cinematic Ontology 193
Josef Früchtl

12. Societies of Control and Chrono-Topologies
M. Christine Boyer 203

13. Digital Architecture and the Temporal Structure of the Internet Experience
Antoine Picon 223

List of Contributors 237

Introduction

Leslie Kavanaugh

The twentieth century has seen many revolutions - historically unprecedented events politically, economically, socially, technologically, artistically. In terms of art, the move to abstraction and non-figurative expression was a revolution that went beyond the effects of photography, cinema, and computer animation technologies. Politically, communism, no matter what one feels about it, was an extraordinary experiment in human governance. Never before had communities tried to cooperate together in this way for mutual benefit of the whole. Of course in the seventeenth century, Europe expanded into a global economy, but the twentieth century saw the mutual interdependence of the world economies, transcending the economic autonomies of nation states. Developments in human thought and possibility also brought many revolutions technologically, of course. The century was one of mobility, and saw the advent of mass migrations of people – not only into new territories, ever-shifting boundaries, but mass mobility of populations on a daily basis just to go to work. Never before in human history have the everyday lives of every man been occupied with getting from point "a" to point "b".

Yet increasingly, we live an uprooted existence, forever wandering waywardly. We seemingly live in a world were we are "connected" to everything, yet have never been more estranged. We are the voyeurs in the lives of peoples we cannot grasp, occupying places on the far side of the globe from us, yet we remain further away than ever because we do not understand other people, and indeed probably cannot comprehend them. In our hubris we think that our presence (even our virtual presence) is innocent. Yet this revolution that has allowed us to peek seditiously into the intimate chambers of our fellow humans – into their countries, into their bodies, into their struggles for human dignity – has not left us feeling more compassionate or connected with the human race as a whole. Just the opposite; this unprecedented access has left us afraid and uncomprehending. And fearful people do violent things.

Consequently, no one can deny that the twentieth century was also the century of unforeseen and unimaginable brutality. The technological scale of wars, not only the two world wars, but innumerable genocides, proved the twentieth century to be quite singular in human history; hundreds of millions of people were slaughtered in the twentieth century. We have never been more close to each other, yet never been so far away from our own humanity.

This collection of essays, specifically, takes up only one revolution that occurred in the twentieth century – the unity of space and time as ontological categories. In 1905, Einstein's paper, *Zur Elektordynamik bewegter Körper*, the so-called Special Theory of Relativity, hit the century like a bomb. Although in popular culture, Einstein is the most well-known theoretical physicist, many scientists contributed to the immense revolution in our conceptions of the physical world and man's place in the universe. Never again could we think of space and time in the same way, even though many theories would rehearse or rehabilitate notions of space and time already postulated in antiquity. Each space and time is specific and individualized.

In a seminar room at the Technical University of Delft, in the Netherlands, a small group of people gathered in order to think through what could space-time mean for the twenty-first century. How to think space-time in an unprecedented way? How to think space-time beyond the disciplines of mathematics and theoretical physics? This community of scholars, moderate in numbers, yet expansive in ambition, gathered once a week with invited guests from all over the world, in order to bend over these questions. The room itself no longer physically exists, destroyed by fire on the 13th of May, 2008. However, the inquiry remains. This volume should be seen as a moment in time, when for several months, we came together with questions circulating about our heads, as Nietzsche would say. These essays are the result of that inquiry.

This volume is organized thematically with regard to the manner in which space-time is approached. Firstly, since the term "space-time" originated with Minkowski, a contemporary of Einstein, Peter Galison jump starts the collection with an excellent explanation of Minkowski's genesis of the concept of space-time. Galison's essay entitled, "Minkowski's Space-Time: From Visual Thinking to the Absolute World", not only sketches out the historical context and development of Minkowski's Space-Time, but also his attempt to provide a unified "Theory of the Absolute World". Notably, for Minkowski, neither Lorentz nor Einstein had truly provided a new conception of space, yet it must be said, in the end, he too wanted a world that was absolute and unchanging.

The second essay by Richard T.W. Arthur entitled, "Materialist Theories of Time", goes back to antiquity in order to unravel time from matter and space. Many theories of a matter-dependant chronology existed long before the twentieth century. What would constitute a materialist theory of time? A first thought might be that, if matter is what exists in space and time, then time should be an existing something within which matter can undergo its changes. Arthur traces the various material theories of time from antiquity, through the modern sciences, and up to Balfour's recent theories supporting the relativist line taken by Leibniz and Mach, albeit incorrectly in

Arthur's view. Leibniz remains an important seventeenth century source whose thought has scarcely been sufficiently explored, specifically in matters concerning a relational theory of space and time.

Other important thinkers, often forgotten, are the mathematicians, scientists and philosophers of the Arabic language who preserved much scholarship of antiquity as well as made strident progress in thinking of space and time. Nader El-Bizri, in his essay entitled, "Corollaries on Space and Time: A Survey of Arabic Sources in Science and Philosophy", traces the origins and development of mediaeval Arabic scholars on the topic of space and time. Whereas notions of place were extended by mathematical notions of geometrical space, time was caught, often, in theological arguments as to the parts of time, and the existence of a future. Obviously, these thinkers are fundamental for understanding the scientific conceptions of space-time inherited in Europe in the seventeenth century, effecting even today the core discussions as to the nature of the physical universe.

Remaining on the larger scale of the universe, Chunglin Kwa takes up the notion of evolutionary time, and space as variation in Darwin, in his essay entitled: "Agency and Space in Darwin's Concept of Variation". In Kwa's discussion of evolutionary time, he positions Darwin's "Natural Selection" as heir to an 18th century mechanist understanding of nature's history. Time would become the relational sequence between events, rather than the neutral background on which timeless laws of nature unfold. Space became the theater in which organisms act out the advantages which variations in their biological constitution had given them. Thus, the notion of "event" with Darwin, gives us not only a "relative" conception of space and time, but also a dynamic, interactive one.

In Leslie Kavanaugh's essay, "The Time of History/The History of Time", the notion of time as history, both collective and individual is explored through several twentieth century historians and philosophers. Through the examination of, for example, Koselleck, Poulantzsas, Gurvitch, and Foucault, time is seen as not something linear, stretching endlessly backwards and forwards through historical time; rather, as spatio-temporal layers. In the considerations of the historical, space and time can also no longer be thought of as separate categories. Foucault's notions of power, albeit problematic, can be used to explain how these spatio-temporal layers are connected. In the end, however, each layer, no matter if it exists upon an epochal timeframe, or an individual one, is a matter of relations, and one that Kavanaugh argues, can be seen as dynamic and generative.

Turning to individual histories and time as memory, Mary Lynne Ellis contributes an essay entitled, "Places Lived in Time", exploring the notions that although our experiences arises from a specific moment in time, they are not fixed and unchanging. Each experience or memory is constituted within a place of their socio-historical and cultural specificity. Ellis provides a post-Freudian theoretical framework, with such thinkers as Fanon, Bulter, Irigaray, and Kristeva, each in their own way providing a critique of the possibility of an "inside", or an "outside". Subsequently, subjectivity can be then thought of as fluid and changing, yet as that which guarantees the continuity of their identity across time. Merleau-Ponty, in turn, theorizes how we are embodied through, and by, the passage of time, including our anticipation of the future. Ellis argues, then, the spatio-temporal context is always present in time as memory.

Patricia Locke in her essay entitled, "Intermittences: Merleau-Ponty and Proust on Time and Grief", also takes up the concepts of embodiment from Merleau-Ponty, but juxtaposed with the literary masterpiece of Marcel Proust, *À la recherche du temps perdu.* Merleau-Ponty philosophically expands upon Husserl's schema that denotes time as "not a line, but a network of intentionalities." We always experience time as a continuous flowing event, as a moving "behind" to what is past. Proust's project of "lost time" mimics affective time, according to Locke, with its flow, its looping back upon itself, its seeming eddies and sudden jerking forward towards the future. Ultimately, Proust's task is to "regain time".

Remaining with the notion of space and time as centered upon the body, upon the experience of the "soul" in Aristotle's terms, Sander van Maas contributes an essay entitled, "Lyrical Bodies: Music and the Extension of the Soul". Where are we when we listen to music? Where is the place of music? The philosophical turn toward the ear, according to van Maas, implies a spatial redistribution and redefinition that ultimately affects the whole of our understanding of the world, the subjects that live in it, and the way world and subjects interrelate. In this essay, Jacques Derrida, referring to the thinking of Jean-Luc Nancy, and Peter Sloterdijk are examined with regard to the space of listening as it unfolds in musicalized, lyrical bodies, which seem to touch one another without apparent contact. These lyrical bodies obey the musical haptics that Plato described when he said that musical harmonies can "take strongest hold upon the soul". With music, we are in a sonorous present.

Another investigation into the spatiality of music is assessed by Raviv Ganchrow in his essay entitled, "Phased Space". From advanced imaging devices to emerging spatial acoustic systems, developments in contemporary technology display a hybridization of sonic and optical attributes. Underpinning these various techniques are complex understandings of spatialized fluid dynamics. This essay traces implicit interdependencies emerging from within the confines of *phased* space, first as they appear in the research of the physicist/philosopher Ernst Mach, and later reflected in the contemporary application of wave field synthesis (WFS). Beginning with Mach's proto-spatial definition of "elements", a position is examined whereby both the "thing" and the discreet "ego" are viewed as alternating pattern-bundles composed of common blocks. The emphasized intertwining of observer and observed along with a broadened definition of "sensation" is then shown to manifest in Mach's empirical research as a radical re-thinking of the epistemic purpose ascribed to the photographic plate. The theme of interdependent tactility and fluctuant spatiality is then applied to a reading of the wave field synthesis technique. This emerging cultural practice involving spatial sound rendering extends the "memory substance" modality into the audio spectrum by implementing a bounded acoustic expanse constructed as a vibrational tapestry, within which a spatialized act of listening occurs divorced from all source resonating "objects".

In addition, a common analogy of the spatio-temporal nature of movement is found in cinema. Josef Früchtl in his essay, "The Evidence of Film and the Presence of the World: Jean-Luc Nancy's Cinematic Ontology", begins with Deleuze's critique of cinema as the "time-image" in classical cinema, where traditionally everything is centred on stability of meaning. Modern cinema as time-image, in contrast, precisely de-stabilizes the possibility of this permanence. Yet Jean-Luc Nancy does not greave this loss; rather, emphasizes its liberation, where the loss of stable meaning also can imply a hybrid meaning, or a state of "not being stable yet". Following Erwin Panofsky, who in his essay "Style and Medium in the Motion Pictures" names as the two specific achievements of film the "dynamisation of space" and the corresponding "spatialisation of time", Kiarostami's films are investigated. Due to the dimension of time, for Nancy, *intensity* as a mode of perception and *immediacy* as a mode of cognition ultimately correspond to cinema's time mode. Film provides a mobile, picturesque experience of time, a mobile experience of space through cinematic movement.

The last two essays deal with time and the city, broadly speaking. Taking up Foucault as a critique of the very structures of power, Christine Boyer in her essay, "Societies of Control and Chrono-Topologies", emphatically makes a case for the spatio-temporal, not as a void, but "where we reside" in a system of relations. This very definition of emplacement enables a reading of a complex of different spaces and separable times. Boyer takes in turn, George Kubler's, *The Shape of Time: Remarks on the History of Things,* Reyner Banham's *Theory and Design in the First Machine Age,* and the cybernetician Gordon Pask's work on communications theory. The so-called arrow of time is not concerned with providing a neat theory of how historical change operated, how works of art were transmitted over time; rather, what is the "shape" of time, with all of its interruptions, variations, and indeterminacies. Not only is the future indeterminate, but time cannot be seen as a formal sequencing that enabled future predictions, that controlled the forward trajectory of thought. The role of historians, Kubler reiterated, is to discover the "manifold shapes of time", not in order to recreate and re-enact these shapes, but to compose meaning from a tradition, to communicate a pattern.

Lastly, in the most contemporary and forward-looking essay of the collection, Antoine Picon provides an essay entitled, "Digital Architecture and the Temporal Structure of the Internet Experience". Picon argues that in many ways, time as "permanence" was the dominant eternal metaphor applied to architectural production. Architecture followed the historical developments in materials, methodologies, and technology. Architecture even resisted the speed of "mechanization", yet the most recent computer software has had a profound effect not only on the production of architecture, but on its "temporality". Notably, Picon does not wish to advance a thesis that architecture is technologically determined; the case is much more complex. Yet increasingly, although architecture has always been about "space"; with the digital revolution, architecture begins to have another relationship to space-time. Although contemporary architecture often employs dynamic images that evoke the days of early photographic images such as by Muybridge or Marrey, this imagery has a paradoxical relationship to historical time, disconnected from any real kind of historical perspective. Architecture is perhaps no longer about "permanence", rather is about "frozen flows", ever-deferred meaning, and a state of suspension from traditional tectonic assumptions.

Conclusively, these twelve papers collected in this volume, explore the consequences of time, and its relationship with space through a multi-disciplinary approach, including the philosophy of space and time, social geography, post-Marxian social theory, new network theory, philosophy of art and culture, musicology, evolutionary biology, historiography, psychoanalytic theory, and comparative literature. The chief inspiration for this collection came from the realization that although mathematics and the quantum-relativity debate in physics had established the concept of space-time, this realization had not been worked through into other areas of concern with concepts of spatiality and temporality. In coming together to "think through" the consequences of this twentieth century revolution in space-time, hybrid spatialities and multiple temporalities became heterogeneous and yet interrelated with the others. Not only can we only now speak of "space-time", and not "space" and "time" as separate ontological categories, but also each becomes multiple. As every existent thing is determined in its own specific location in time; we can only speak of "times" and of "spaces" in the particular plural. This determination effectively implies not only the ontological status of every sensible object, but the fact that every existent thing is precisely individuated by its singular position in space-time. Yet one idea has not changed, one spatio-temporal orientation - the possibility of a different future, the hope, ever-abiding, for a constructive change.

Minkowski's Space-Time: From Visual Thinking to the Absolute World*

Peter Galison

Galison traces Minkowski's progression from his visual-geometric thinking to his physics of space-time and finally to his view of the nature of physical reality. Minkowski always held that a sort of "pre-established harmony" existed between mathematics and nature, but then a different sort of "pre-established harmony" than that of Leibniz. Geometry was not merely an abstraction from natural phenomena or a mere description of physical laws through a mathematical construct; rather, the world was indeed a "four-dimensional, non-Euclidean manifold", a true geometrical structure. As a contemporary of Einstein, Minkowski proposed a reconciliation of gravitation and electro-magnetism that he called, "the Theory of the Absolute World". Despite his untimely death, Minkowski holds a prominent place in twentieth century theoretical physics, not the least for his conception of "space-time", emphatically stating that we can no longer speak of "space" and "time", rather "spaces" and "times".

Hermann Minkowski is best known for his invention of the concept of space-time. For the last seventy years, this idea has found application in physics from electromagnetism to black holes. But for Minkowski space-time came to signify much more than a useful tool of physics. It was, he thought, the core of a new view of nature which he dubbed the "Theory of the Absolute World." This essay will focus on two related questions: how did Minkowski arrive at his idea of space-time, and how did he progress from space-time to his new concept of physical reality.[1]

---

* This essay is reprinted with permissions from the author and publisher from *Historical Studies in the Physical Sciences*, Vol.10, (1979):85-119. Originally this periodical was published by Johns Hopkins University, but is now held by the University of California Press, ISBN/ISSN: 00732672. Permissions to republish granted through Copyright Clearance Center, Confirmation Number: 1792389.

[1] A different approach to Minkowski's work has been taken by Lewis Pyenson in "La Réception de la relativité généralisée: disdplinarité et institutionalisation en physique": *Revue d'histoire des sciences*, (28 January, 1975):61-73. Pyenson touches on Minkowski's role in the institutionalization of mathematical physics. Pyenson's "Hermann Minkowski and Einstein's Special Theory of Relativity," *Archive for History of Exact Sciences,* 17 (1977):71-95, deals more closely with Minkowski's space-time in special relativity, in order to link Minkowski's work with the institutional history of mathematical physics. Among other sources from the manuscripts at the American Institute of Physics, Pyenson uses the first ten pages of the "*Funktionentheorie*" lecture discussed in this essay and he provides a valuable transcription of this lecture. The contrast between Minkowski's mathematical physics and the theoretical physics of Einstein and others is further discussed both in Russell McConnmach's "Editors Forward" to *Historical Studies in the Physical Sciences,* 7 (Philadelphia: University of Pennsylvania Press, 1976), pp, xi-xxxv, and in Pyenson's article, "Einstein's Early Sdentific Collaboration", *Historical Studies in the Physical Sciences,* 7, pp.83-123.

Minkowski was born on 22 June 1864 in Alexotas, Russia. In 1872 his family immigrated to Prussia, where Minkowski attended the Altstädtisches Gymnasium in Königsberg. Like all students of the Gymnasium, he received a strict education which stressed both classical literature and science. Minkowski graduated early and went on to the University of Königsberg, where he studied principally with Wilhelm Weber and Woldemar Voigt. He then spent three semesters in Bonn, returning to Königsberg to receive his doctorate in 1885. As was the custom, he spent several years teaching as a privatdozent in Bonn, where he remained until 1894. After two more years at Königsberg he moved to Zurich for six years. Then, in 1902, at David Hilbert's request, a chair was created for him in Göttingen, where he remained until his untimely death in 1909.[2]

Minkowski's extraordinary geometric insight was evident by the time he was seventeen, when he won the grand prize of the French Academy for a geometric, general treatment of the theory of quadratic forms.[3] Contemporary mathematicians immediately recognized his talent. On receiving his handwritten manuscript of the prize essay, C. Jordan wrote to young Minkowski, "Please work to become an eminent geometrician."[4] Some years later in 1896, when Minkowski's major work on number theory, *The Geometry of Numbers*, appeared, Hermite wrote to Laugel, "I think I see the promised land."[5] Harris Hancock put it slightly differently, but no less grandly, in the introduction to his *Development of the Minkowski Geometry of Numbers*: "His grasp of geometrical concepts seemed almost superhuman."[6] This "almost superhuman" grasp of geometry was put to work in most of Minkowski's mathematical discoveries. One area of application was "geometry of numbers," where he used geometric methods to derive estimates for positive definite ternary forms. Another was to the theory of continued fractions based on the closest packing of spheres. These researches led him to a detailed study of convex bodies which in turn yielded a host of new number-theoretic advances.[7]

Minkowski spoke explicitly about his use of geometrical insight in an unpublished introductory lecture to a number theory course, dated 28 October 1897:

---

[2] For biographical details of Minkowski see Minkowski, Hennann, *Hermann Minkowski. Briefe an David Hilbert,* eds. L. Rüdenberg and H. Zassenhaus (Berlin: Springer-Verlag, 1973),pp. 9-16.
[3] Hancock, Harris, *Development of the Minkowski Geometry of Numbers* (New York: Macmillan, 1973), p. viii.
[4] Hilbert, David, "Gedächtnisrede, gehalten in der öffentlichen Sitzung der Kgl. Gesellschaft der Wissenschaften zu Göttingen am 1. Mal 1909," *Gesammelte Abhandlungen von Hermann Minkowski* (New York: Chelsea, 1967), p. vii. (author's translation).
[5] Hancock, *Minkowski Geometry of Numbers*, p. viii (author's translation).
[6] *ibid*, p. vii.
[7] Dieudonné, J., "Hermann Minkowski," *Dictionary of Scientific Biography,* 9 (New York: Scribner, 1975), pp. 411-414 (hereinafter *DSB).*

> In [applied number theory] one can frequently make use of geometrical intuition [*geometrischer Anschauung*] for the easier discovery of theorems, and so there arises a field, specific areas of which were first created by Gauss, Dirichlet, Eisenstein, and Hermite, and to which I gave the name Geometry of Numbers. It is therefore essentially a question of using a spatial intuition [*räumlicher Anschauung*] for the uncovering of relations among integers.[8]

Thus geometrical and spatial *Anschauung* form the core of Minkowski's approach to number theory. The distinctive visual cast of the word *Anschauung* (view, outlook, or intuition) is reflected in its scientific usage, both by Minkowski and others, as authors such as Gerald Holton have pointed out.[9] In this context *geometrische Anschauung* might best be translated as "visual-geometric intuition." Minkowski's claim, then, is that through this visual-geometric intuition we will discover the theorems and relations of number theory. In short, Minkowski's mathematical work was characterized by a self-conscious and successful application of geometrical thinking to fields of mathematics outside of geometry proper.

But Minkowski's interests were not confined to geometry. After receiving his doctorate, he spent five years teaching in Bonn. There he became interested in a variety of problems in physics, and in 1888 published an article on hydrodynamics, which was submitted by Hermann von Helmholtz.[10] He continued in the following years to pursue his interest in physics and reported to Hilbert in 1890 that he was learning both about practical and experimental physics and the work of Helmholtz, J. J. Thomson, and Heinrich Hertz.[11] Minkowski's interest in Hertz's work was so strong that he once remarked that had Hertz lived, he (Minkowski) might have turned more completely from mathematics to physics.[12]

---

[8] "Zu ihr [die angewandte Zahlentheorie] kann man vielfach von geometrischer Anschauung zur leichteren Auffindung von Sätzen Gebrauch machen und so entsteht ein Gebiet, welches zuerst in einzelnen Partien bei Gauss, Dirichlet, Eisenstein, Hermite auftaucht und welchem ich den Namen Geometrie der Zahlen gegeben habe. Es handelt sich von demselben also wesentlich um einen Gebrauch räumlicher Anschauung zur Aufdeckung von Beziehungen für ganze Zahlen ..." Minkowski, Hermann, unpublished lecture, 28 October 1897, *MP,* box V, folder 13. See appendix "Notes on Manuscript Sources".

[9] Holton, Gerald, *Thematic Origins of Scientific Thought: From Kepler to Einstein* (Cambridge, Massachusetts: Harvard University Press, 1973), p. 370, and Arnheim, Rudolf, *Visual Thinking* (Berkeley: University of California Press, 1969), p. 299, discuss *Anschauung* in relation to Pestalozzi's pedagogical use of visual intuition. Holton's analysis of Einstein's visual way of thinking is discussed further in the conclusion of this essay.

[10] "Ueber die Bewegung eines festen Körpers in einer Flüssigkeit," *Sitzungsberichte der Berliner Akademie* 1888, pp. 1095-1110, cited in Hilbert, "Gedächtnisrede," p. xxi.

[11] Minkowski, *Briefe an David Hilbert*, pp. 39-40.

[12] Hilbert, "Gedächtnisrede," pp. xxi-xxii.

Minkowski arrived in Göttingen in 1902 during a period of great excitement in the physics community. By 1900, H. A. Lorentz had achieved remarkable success in explaining both the optical and the electrodynamic properties of objects moving in the ether using only electromagnetic considerations.[13] Wilhelm Wien had responded to these triumphs with a call for a new program for a unified physics. His goal was to bring together mechanics and electrodynamics by explaining mass purely electrodynamically. "It is doubtless one of the most important tasks of theoretical physics," Wien wrote, "to unite the two heretofore completely isolated fields of mechanical and electrodynamic phenomena, and to derive from a common foundation their respective differential equations."[14] Lorentz's 1904 paper caused further excitement by rendering Maxwell's equations covariant to all orders of *v/c* and thus "explaining" the puzzling Michelson-Morley experiment.[15]

At Göttingen these developments were closely watched, for between 1900 and 1910 there were more physicists working on electron theories there than at any other university.[16] Since the time of C. F. Gauss and Wilhelm Weber, Göttingen had been a center for electromagnetic research.[17] In the years following their pioneering work, Eduard Riecke had worked on the electrical properties of metals,[18] and Emil Wiechert,[19] Carl Runge,[20] and Arnold Sommerfeld[21] had conducted further research in electron theory. During Minkowski's career at Göttingen, Max Abraham[22] and Walter Kaufmann[23] moved to the forefront of research on the dynamics of the electron. Minkowski himself joined Hilbert in conducting several seminars on the new research in electrodynamics. According to both Hilbert and Max

---

[13] On Lorentz's work see McCormmach, Russell, "H. A. Lorentz and the Electromagnetic View of Nature," *Isis,* 61 (1970): 459-497; Schaffner, Kenneth, "The Lorentz Theory of Relativity," *American Journal of Physics*, 37 (1969): 498-513; and Miller, Arthur I., "On Lorentz's Methodology," *British Journal for the Philosophy of Science, 25* (1974): 33*ff.*

[14] Wien, Wilhelm, "Ueber die Möglichkeit einer elektromagnetischen Begründung der Mechanik," reprinted in *Annalen der Physik,* 5 (1901): 501.

[15] Lorentz, H. A., "Electromagnetic Phenomena in a System Moving with Any Velocity Smaller than that of Light," *Proceedings of the Royal Academy of Amsterdam*, 6 (1904): 809, reprinted in H. A. Lorentz, A. Einstein, H. Minkowski, and H. Weyt, *The Principle of Relativity: A Collection of Original Memoirs on the Special and General Theory of Relativity,* trans. W. Perett and G. B. Jeffrey, nn. by A. Sommerfeld (New York: Dover, 1952), pp. 1-34. See note 25.

[16] Lewis Pyenson, "Einstein's Early Collaboration," *Historical Studies in the Physical Sciences,* 7 (1976): 89.

[17] Woodruff, A. E., "Wilhelm Eduard Weber," *DSB,* 14, pp.203-209.

[18] Goldberg, Stanley, "Eduard Riecke," *DSB,* 11, pp.445-447.

[19] Bullen, K. E., "Emil Wiechert," *DSB,* 14, pp.327-328.

[20] Forman, Paul, "Carl David Tolme Runge," *DSB,* 11, pp.610-614.

[21] Forman, Paul and Hermann, Armin, "Arnold Sommerfeld," *DSB*, 12, pp.525-532

[22] Goldberg. Stanley, "Max Abraham," *DSB,* 1, pp.23-25. For more on Abraham's work see Goldberg, "The Abraham Theory of the Electron: The Symbiosis of Experiment and Theory," *Archive for History of Exact Sciences,* 7 (1970): 7-25.

[23] Campbell, John T., "Walter Kaufmann," *DSB,* 7, pp.263-265.

Born it was during these seminars that Minkowski first began to develop his new ideas on relativity and space-time.[24]

The purpose of this paper is to trace Minkowski's progression from his visual-geometric thinking to his physics of space-time, and finally to his view of the nature of physical reality. He held that because of a "pre-established harmony between mathematics and nature", geometry could be used as a key to physical insight. Thus he was able to justify relativity as the physical theory with the more satisfying geometrical structure. But this mathematical structure came to mean more to Minkowski than simply a reformulation of relativity. Minkowski eventually ascribed physical reality to the geometry of space-time.

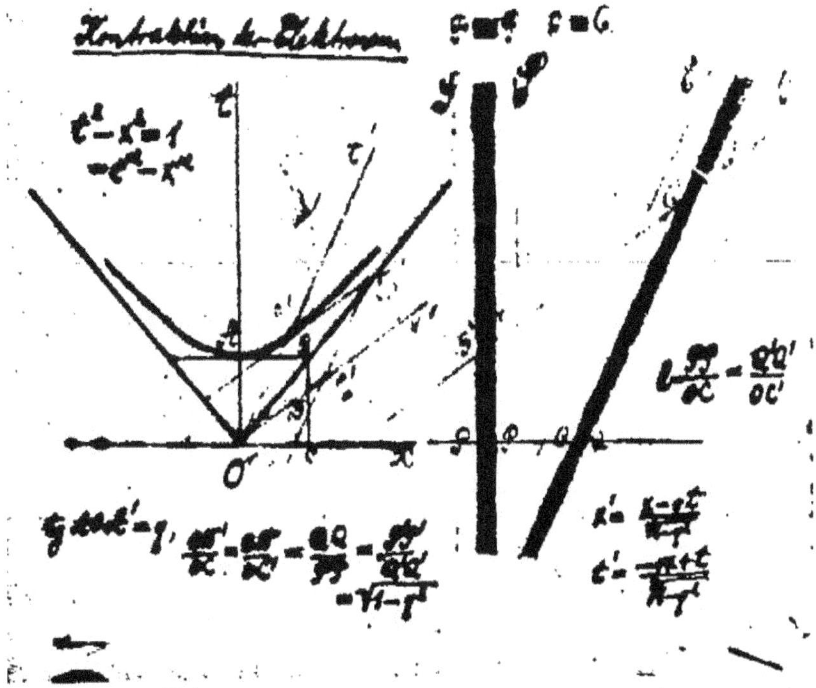

Figure 1: Lecture Slide
Photocopy of slide presented in the lecture "Space and Time", Cologne, 21 September 1908, before the *Versammlung Deutscher Naturforscher und Ärzte.*

For sources on Minkowski's development of space-time we can draw on three public presentations of his views: the speeches "Space and Time" and "The Principle of Relativity", and the paper "The Fundamental Equations of

[24] Born, Max, "Erinnerungen an Hermann Minkowski zur 50. Wiederkehr seines Todestages," *Die Naturwissenschaften,* 17 (1959): 502; Hilbert, "Gedächtnisrede," p. xxii.

Electromagnetic Phenomena in Moving Bodies." "Space and Time" ("*Raum and Zeit*") was delivered to the eightieth meeting of the Assembly of Natural Scientists and Physicians in Cologne on 21 September 1908. The famous space-time diagram was first seen as a lecture slide (reproduced in Figure 1) and perhaps sketched for the first time in Minkowski's notes for the talks (Figure 2). The lecture was first printed in 1909 in the *Physikalische Zeitschrift*.[25] "The Principle of Relativity" ("*Das Relativitätsprinzip"*) was presented to the Göttingen Mathematische Gesellschaft on 5 November 1907, and published by Arnold Sommerfeld in 1915 in the *Annalen der Physik*, six years after Minkowski's death.[26] The ideas on space and time developed in these lectures were, to some extent, applied in a major work on the laws of electrodynamics, "The Fundamental Equations for Electromagnetic Phenomena in Moving Bodies" *("Die Grundgleichungen für die elektromagnetische Vorgänge in bewegten Körpern*"), published in 1908.[27] The "*Grundgleichungen*" is important for both its results (the first relativistically correct presentation of Maxwell's equations in a ponderable medium), and its mathematical formalism (tensor calculus). In addition, I have found a variety of unpublished papers relevant to Minkowski's views on space and time. These are used throughout the paper and are described in the appendix, "Notes on Manuscript Sources."

Minkowski and the Electromagnetic World Picture

As has been amply discussed in the literature, it was Einstein's contribution to have abandoned the search for a dynamics of the electron by turning first to the kinematics of macroscopic bodies.[28] This involved an epistemological criticism of the concepts of space and time. By provisionally neglecting the goal of finding a dynamics of the structure of matter, he arrived deductively

[25] *Physikalische Zeitschrift,* 10 (1909): 104-111, reprinted in H. A. Lorentz *et al., Das Relativitätsprinzip,* 5th ed. (Stuttgart: B. G. Teubner, 1974), pp. 54-56, hereinafter RZ. This edition has been translated as *The Principle of Relativity* (see note 15), hereinafter *ST* (for "Standard Translation"); the article "Space and Time" in this edition is the standard translation of "*Raum und Zeit*." References will be first to the German and then to the English, *e.g.,* RZ, p. 54; ST, p. 75. English translations are from ST except where 947 sciences-10: 58.

[26] *Annalen der Physik,* 47 (1915): 927-938 (hereinafter RP). Translations from RP are the author's.

[27] *Nachrichten der Königlichen Gesellschaft der Wissenschaft und der Georg-August Universität zu Göttingen, Mathematisch-physikalische Klasse* (1908), pp. 53-111, reprinted In Hilbert, *Abhandlungen von Minkowski*, pp. 352-404.

[28] On the Electromagnetic World Picture see Holton, *Thematic Origins*, pp. 177-179; Schaffner, K. F., "Lorentz Theory," *op. cit.* (note 13), esp. pp, 508-513; McConnmach, Russell, "Einstein, Lorentz and the Electron Theory," *Historical Studies in the Physical Sciences,* 2 (1970): 69-81; Miller, Arthur I., "A Study of Henri Poincaré's 'Sur la Dynamique de l'électron'," *Archive for History of Exact Sciences,* 10 (1973): 207-328, esp. 207-233; Miller, Arthur I., "The Physics of Einstein's Relativity Paper of 1905 and the Electromagnetic World Picture of 1905" *American Journal of Physics;* Hirosige, Tetu, "Theory of Relativity and the Ether," *Japanese Studies in the History of Science,* 7 (1968): 37-58; and the items in note 13.

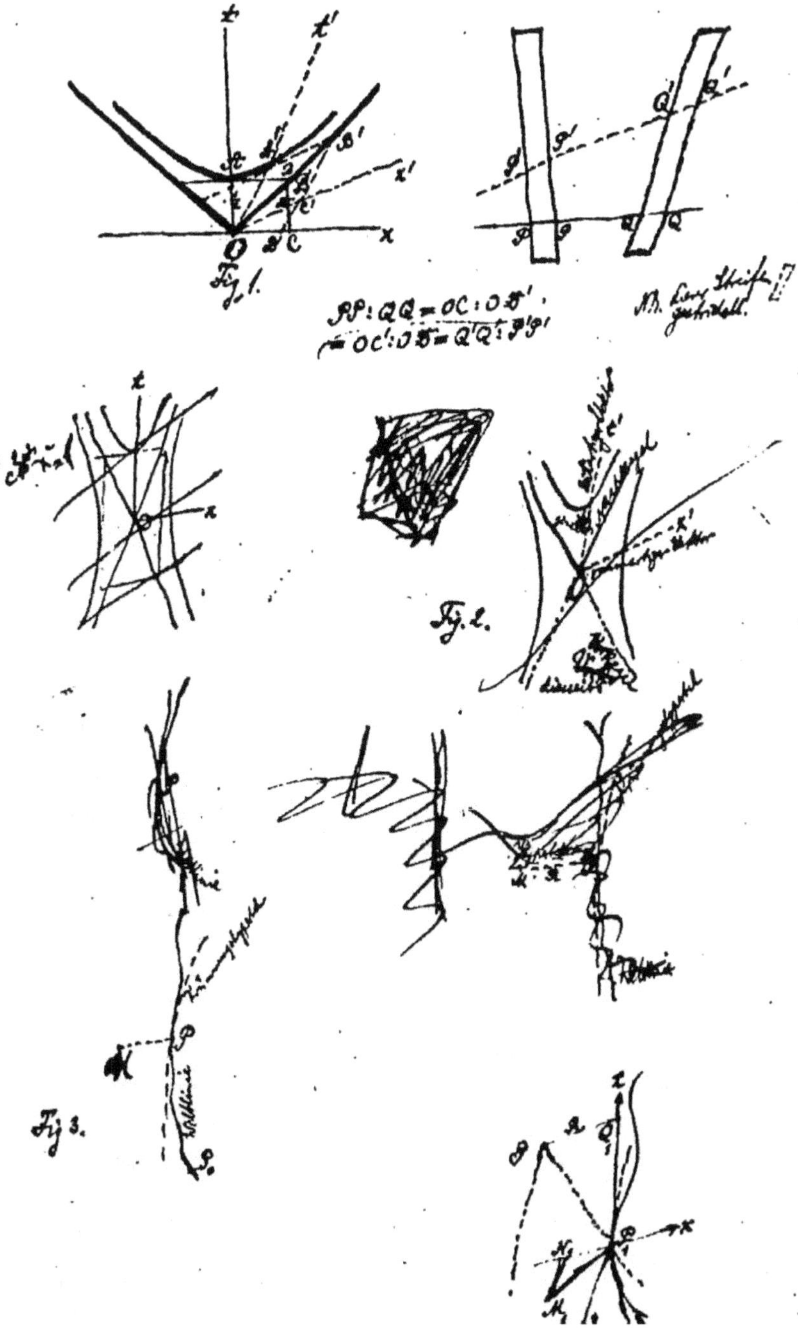

Figure 2: Early Space-Time Diagrams.
These may be the first space-time diagrams ever drawn. From the Göttingen Archives Draft RZ 2, p. 10, unlabeled.

at the space and time transformations which Lorentz had adopted without justification. Conversely, Einstein's explanation of the null result of the optical experiments followed directly from his axioms, while Lorentz and Henri Poincaré had to explain these results with hypotheses on matter and forces. The Significance of Einstein's work was not, however, immediately clear to physicists of the day. While Einstein's theory of special relativity was published in 1905, it had just begun to gain acceptance in 1910.[29] Instead, physicists like Poincaré, Lorentz, and Abraham continued to work towards the goal of explaining all of mechanics purely electro-dynamically in accordance with the reductionist view which held the electron to be the fundamental building block of all matter. In fact, all three of these scientists continued to search for an electromagnetic explanation of mass until their respective deaths.

Minkowski's view on the question of reductionism is apparent in "The Principle of Relativity," where he writes, "Here we find ourselves at a standpoint where the true physical laws are not yet completely known to us. One day, perhaps, a reduction will be possible based on purely electrical consideration....,"[30] or again in the draft of the introductory lecture on complex analysis, where he remarks, "Physicists suspect that one day it might be possible to explain all natural phenomena purely electrodynamically so that finally there will be no matter, nothing but electricity in the world."[31] Assertions such as these make it clear that Minkowski did not break with tradition in order to follow Einstein's new approach. In fact, when Minkowski refers to Einstein, it is evident that he saw Einstein as furthering the Electromagnetic World Picture. He asserts, for example, that "from this very strange sounding hypothesis [the Lorentz contraction hypothesis], the Postulate of Relativity was finally developed in a form which is exceptionally accessible to the mathematician. We owe the working out of the general principle to Einstein, Poincaré and Planck, whose work I will shortly consider more closely."[32] This evaluation of Einstein's role is voiced again in an unpublished draft of "The Relativity Principle": "As to the merits of the individual authors: the (essential) foundation of the ideas originates with Lorentz; Einstein more cleanly developed the Principle of Relativity. At the

---

[29] Holton, *Thematic Origins*, pp. 268-269.

[30] "Hier stellen wir uns auf den Standpunkt, die zutreffenden physikalischen Gesetze sind uns noch nicht völlig bekannt. Eines Tages würde vielleicht eine Zurückführung auf reine Elektrizitätslehre möglich sein..." *RP*, *op. cit.*, p. 931.

[31] "Nun schwebt den Physikern der Gedanke vor, dass es eines Tages gelingen möchte, alle Naturvorgänge rein elektrodynamisch zu erklären, so dass es schliesslich auf der Welt nichts anderes als Elektrizität, keine Materiel gieht." *Lecture on complex analysis*, pp. 11-12. See appendix "Notes on Manuscript Sources" for details.

[32] " Aus dieser höchst seltsam klingenden Hypothese hat sich dann schliesslich das Postulat der Relativität in einer Form herausentwickelt, die dem Verständnis des Mathematikers besonders gut zugänglich ist. Verdienste um die Ausarbeitung des allgemeinen Prinzips haben Einstein, Poincaré und Planck, über deren Arbeiten ich alsbald Näheres sagen werde." RP, *op. cit.*, p. 928.

same time he applied it with special success to the treatment of special problems of the optics of moving media, and finally he was also the first to draw the consequences of the variability of mechanical mass in thermodynamic processes." [33] Minkowski, like Lorentz, took the transformations to be an explanation of phenomena, whereas Einstein considered them as a consequence of our measurement of space and time. It is thus understandable that Einstein's work appeared to Minkowski as a generalization of Lorentz's. Minkowski made no mention, to this point, of Einstein's critical contribution to the understanding of the physical significance of the transformations.

With this background, one is not surprised at Minkowski's assessment of his own contribution to relativity theory, which he refers to as the "World Postulate" (to be discussed below). This judgment occurs at the end of "Space and Time," where Minkowski returns to mechanics to present it in "harmony" with electrodynamics: "The validity without exception of the world-postulate, I like to think, is the true nucleus of an electromagnetic image of the world, which, discovered by Lorentz, and further revealed by Einstein, now lies open in the full light of day."[34] In sum, Minkowski still hoped for the completion of the Electromagnetic World Picture through relativity theory. Moreover, he saw his own work as completing the program of Lorentz, Einstein, Planck, and Poincaré. Of these, it was Poincaré who most directly influenced the mathematics of Minkowski's space-time.

As Minkowski acknowledges many times in "The Principle of Relativity", his concept of space-time owes a great deal to Poincaré's work.[35] "*Sur la Dynamique de l'électron*" contains Poincaré's systematic search for the invariants of the Lorentz transformation. But of even greater significance

---

[33] "Was das Verdienst der einzelnen Autoren angeht, so rühren die {wesentlichen} Grundlagen der Ideen von Lorentz her, Einstein hat das Prinzip der Relativität reinlicher herauspräpariert, zugleich es mit besonderem Erfolge zur Behandlung spezieller Probleme der Optik bewegter Medien angewandt, endlich auch zuerst Folgerungen über Veränderlichkeit der mechanischen Masse bei thermodynamischen Vorgängen gezogen." Draft RP A, p. 16. See appendix "Notes on Manuscript Sources" for details. Crossed brackets indicate that the enclosed work was crossed out in the original. At this early time, (1907) it is clear that Minkowski did not understand the import of Einstein's theory. It is therefore surprising to read in Max Born's *Autobiography* (New York: Charles Scribner's Sons, 1978), p. 131, that "[Minkowski] told me later that it came to him as a great shock when Einstein published his paper in which the equivalence of the different local times of observers was pronounced; for he had reached the same conclusions independently but did not wish to publish them because he wished first to work out the mathematical structure in all its splendour." My thanks to Professor I. B. Cohen for showing me this quotation.

[34] "Die ausnahmslose Gültigkeit des Weltpostulates ist, so möchte ich glauben, der wahre Kern eines elektromagnetischen Weltbildes, der von Lorentz getroffen, von Einstein weiter herausgeschält, nachgerade vollends am Tage liegt." *op. cit,* RZ, p. 66; ST, p. 91.

[35] Poincaré, Henri A., "Sur la Dynamique de l'électron", *Rendiconti del Circolo Matemalico di Palemlo,* 21 (1906): 129-175; partial translation in W. Kilmister, ed.; *Special Theory of Relativity* (Oxford: Pergamon, 1970), pp. 144-185. For a critical work on this paper see Miller, "A Study of Poincaré's 'Sur la Dynamique'," *op. cit.*

to Minkowski was Poincaré's four-dimensional *x, y, z, ict*-space, which contains the seeds of the four vector formalism. If, Poincaré writes,

$$\begin{matrix} x & y & z & t\sqrt{-1} \\ \delta x & \delta y & \delta z & \delta t\sqrt{-1} \\ \delta_1 x & \delta_1 y & \delta_1 z & \delta_1 t\sqrt{-1} \end{matrix}$$

> are regarded as the coordinates of three points *P, P', P"* in four-dimensional space, we see that the Lorentz transformation is simply a rotation of this space about a fixed origin. The only distinct invariants are therefore the six distances of the points *P, P', P"* from one another and from the origin, or alternatively the two expressions
>
> $$x^2 + y^2 + z^2 - t^2, \qquad x\delta x + y\delta y + z\delta z - t\delta t$$
>
> and the four expressions of the same form obtained by permuting the three points *P, P', P"* in any manner.[36]

Minkowski clearly draws on the interpretation both of the invariants as distances and of the Lorentz transformations as rotations in the *x, y, z, ict*-space. Notice that by giving the fourth coordinate the dimensions of *ict,* Poincaré, in contrast to Minkowski, does not emphasize the non-Euclidean nature of the space.

From these invariants Poincaré was able to construct a covariant law of gravitation consistent with special relativity. But Poincaré ascribed neither metaphysical nor physical importance to the four dimensional representation. Indeed, as late as 1908 he asserted:

> It seems in fact that it would be possible to translate our physics into the language of geometry of four dimensions; to attempt this translation would be to take great pains for little profit, and I shall confine myself to citing the mechanics of Hertz where we have something analogous. However, it seems that the translation would always be less simple than the text, and that it would always have the air of a translation, that the language of three dimensions seems the better fitted to our description of the world although this description can be rigorously made in another idiom.[37]

---

[36] Poincaré, "Sur la Dynamique," in Kilmister, *op. cit.*, pp. 175-76.

[37] Poincaré, Henri, *Science and Method,* trans. John W. Boulduc (New York: Dover, 1959), p. 427.

By contrast Minkowski, beginning with precisely the same formalism, came to believe that the laws of physics would only be fully understood in four-dimensional space-time.

The Union of Space and Time

To understand Minkowski's idea of space-time, it is revealing to compare and contrast the introductions to his two lectures, "The Principle of Relativity" and "Space and Time." "The Principle of Relativity" begins with a discussion of the new notions of space and time which follow from relativity theory: "Out of the electromagnetic theory of light there recently seems to have come a complete transformation of our representations [*Vorstellungen*] of space and time which must be of exceptional interest for the mathematician to learn."[38] Minkowski's use of the term *Vorstellung* announces the orientation of his discussion. *Vorstellung* is an abstract term of concrete origin. Literally a "placing before," it has a more substantial connotation than its English translation "representation," "idea," or "conception." Aware of Minkowski's background, we can see his "spatial intuition" used from the outset. More explicitly, Minkowski next stresses the mathematical interest of the new theory:

> The mathematician is also especially well prepared to pick up the new views [*Anschauungen*] of space and time because it involves acclimating himself to conceptual schemes [*Begriffsbildungen*] with which he has long been familiar. The physicist, on the other hand, must discover afresh these concepts [*Begriffe*], and must painfully cut his way through a jungle of obscurities. Meanwhile close by, the old, excellently laid out path of the mathematician comfortably leads forward.[39]

Where the mathematician uses views (*Anschauungen*) and conceptual schemes (*Begriffsbildungen*), the physicist must struggle with concepts (*Begriffe*). The distinct contrast between the visual terms employed to describe mathematics and the more formal terminology used for physics indicates that Minkowski attaches particular significance to his geometrical approach to relativity: "Above all, the new formulation would be, if in fact it

---

[38] "Von der elektromagnetischen Lichttheorie ausgehend, scheint sich in der jüngsten Zeit eine vollkommene Wandlung unserer Vorstellungen von Raum und Zeit vollziehen zu wollen, die kennen zu lernen für den Mathematiker jedenfalls von ganz besonderem Interesse sein muss." RP, *op. cit.*, p. 927.

[39] "Auch ist er besonders gut prädisponiert, die neuen Anschauungen aufzunehmen, weil es sich dabei um eine Akklimatisierung an Begriffssbildungen handelt, die dem Mathematiker längst äusserst geläufig sind, während die Physiker jetzt diese Begriffe zum Teil neu erfinden und sich durch einen Urwald von Unklarheiten mühevoll einen Pfad durchholzen müssen, indessen ganz in def Nähe die längst vortrefflich angelegte Strasse der Mathematiker bequem vonvärts führt." RP, *op. cit.*, p. 927.

correctly reflects the phenomena, practically the greatest triumph applied mathematics has ever shown. Expressed as briefly as possible, it is this - the world in space and time in a certain sense is a four-dimensional, non-Euclidean manifold"[40] Thus Minkowski introduces the four-dimensional space before discussing Einstein's requirement that electrodynamic theory yield a symmetry with respect to transformation of inertial reference systems. Instead of emphasizing the physical basis of relativity, Minkowski stresses the underlying mathematics of geometry. For Minkowski it is not that physical laws can be equivalently expressed through a mathematical construct, but rather that "in a certain sense...the world *is* a four-dimensional, non-Euclidean manifold" (emphasis added).

This being the case, it is the mathematician who is in a position to say something fundamental about reality, rather than the physicist. In short, Minkowski assigns two different but complementary roles to mathematics. First, mathematics offers physics a set of geometric concepts useful in approaching the new relativity theory. Second, Minkowski identifies physical reality, at least "in a certain sense" with mathematical structure. Should the mathematical statements about reality be valid, "it would be revealed, to the fame of the mathematician and to the boundless astonishment of the rest of mankind, that mathematicians, purely in their imagination, have created a large field to which one day the fullest real existence should be ascribed (though it was never the intention of these idealistic fellows)."[41] Here again, he takes mathematics to be reality rather than an abstraction from or an idealization of reality. If the world is a mathematical structure, then the physical-geometrical laws which describe it acquire an ontological status.

The first paragraph of "The Principle of Relativity," which begins with the first citation in this section and ends with the one above, introduces Minkowski's visual-geometric way of thinking. It stresses the importance he places on the mathematics of relativity and on his view that physical reality lies in four dimensions. These convictions are pointedly reemphasized in Minkowski's introduction to his second and most famous lecture, entitled "Space and Time." That lecture begins, however, with an unexpected tribute to the importance of experiment. "Gentlemen! The views of space and time which I wish to lay before you," he writes, "have sprung from the soil of

---

[40] "Überhaupt würden die neuen Ansätze, falls sie tatsächlich die Erscheinungen richtig wiedergehen, fast den grössten Triumph bedeuten, den je die Anwendung der Mathematik gezeitigt hat. Es handelt sich, so kurz wje möglich ausgedrückt - Genaueres werde ich alsbald ausführen - darum, dass die Welt in Raum und Zeit in gewissem Sinne eine vierdimensionale nichteuklidische Mannigfaltigkeit ist." RP, *op. cit.*, p. 927.

[41] "Es würde zum Ruhme der Mathematiker, zum grenzenlosen Erstaunen der übrigen Menschheit offenbar werden, dass die Mathematiker rein in ihrer Phantasie ein grosses Gebiet geschaffen haben, dem, ohne dass dieses je in der Absicht dieser so idealen Gesellen gelegen hätte, eines Tages die vollendetste reale Existenz zukommen sollte." RP, *op. cit.*, pp. 927-928.

experimental physics, and therein lies their strength. They are radical."[42] Here Minkowski seems to feel obliged to defend his departure from the physics of his time which dealt with specific experimental data. The prominent place Minkowski gives experimental physics is surprising, since it is the only reference to experiment in the speech; he may have felt obligated to acknowledge the role of experiment at least once to avoid the charge of being overly speculative.

The next sentence contains, without doubt, Minkowski's most memorable remark: "Henceforth space by itself, and time by itself, are doomed to fade away into mere shadows, and only a kind of union of the two will preserve independence."[43] This sentence in its draft version reads: "Then, from now on, space for itself and time for itself should sink completely into shadows. Only a concept [*Begriff*] obtained by a fusion of the two [concepts] will show a free existence. I will provisionally call this the absolute (?) world until a clear and distinct label for this notion can be found."[44] The published version refers to the "independence" of space-time, the draft form to its "free existence." Both paragraphs emphasize that Minkowski's new space-time exists independently of observer, unlike the old concepts of space and time. It is this freedom from reference frame which entitles the four-dimensional space to be called "The Absolute World."[45]

The draft manuscripts of the "Space and Time" lecture reveal that composing this crucial first page was a struggle for Minkowski; words are crossed out two or three times, phrases are eliminated, replaced, and struck out again (see Figure 3). The terms used in this introduction were very carefully chosen and indicate the importance Minkowski attached to aspects other than the formal results of his investigations. Consider the published sentence, "*Ihre Tendenz ist eine radikale*." The choice of words would seem

---

[42] "M. H.! Die Anschauungen über Raum und Zeit, die ich Ihnen entwickeln möchte, sind auf experimentell-physikalischem Boden erwachsen. Darin liegt ihre Starke. Ihre Tendenz ist eine radikale." *op. cit*; RZ, p. 54; ST, p. 75.

[43] "Von Stund an sollen Raum für sich und Zeit für sich völlig zu Schatten herabsinken und nur noch eine Art Union der beiden soll Selbständigkeit bewahren." *op. cit*., RZ, p. 54; ST, p. 75.

[44] "Dann, von Stund an, sollen Raum für sich und Zeit für sich zu völligen Schatten herabsinken, nur noch ein durch Verschmelzung der Beiden gewonnener Begriff, den ich provisorisch bis zur Ersinnung eines eigenartigeren den Namen absolute (?) Welt geben will, wird eine freie Existenz zeigen." Draft RZ 2, p. 1, labeled 1 (Figure 3).

[45] The standard translation may have led to some misunderstanding on this point because it renders the ending of the above citation as, "...only a kind of union of the two will preserve an *'independent reality'*" (emphasis added); *Selbständigkeit* should be simply "independence." Hans Reichenbach has commented on this confusion: "The first part of Minkowski's remark has unfortunately caused the erroneous impression that all visualizations of time as time and space as space must disappear." Reichenbach, Hans, *The Philosophy of Space and Time* (New York: Dover, 1957), p. 160.

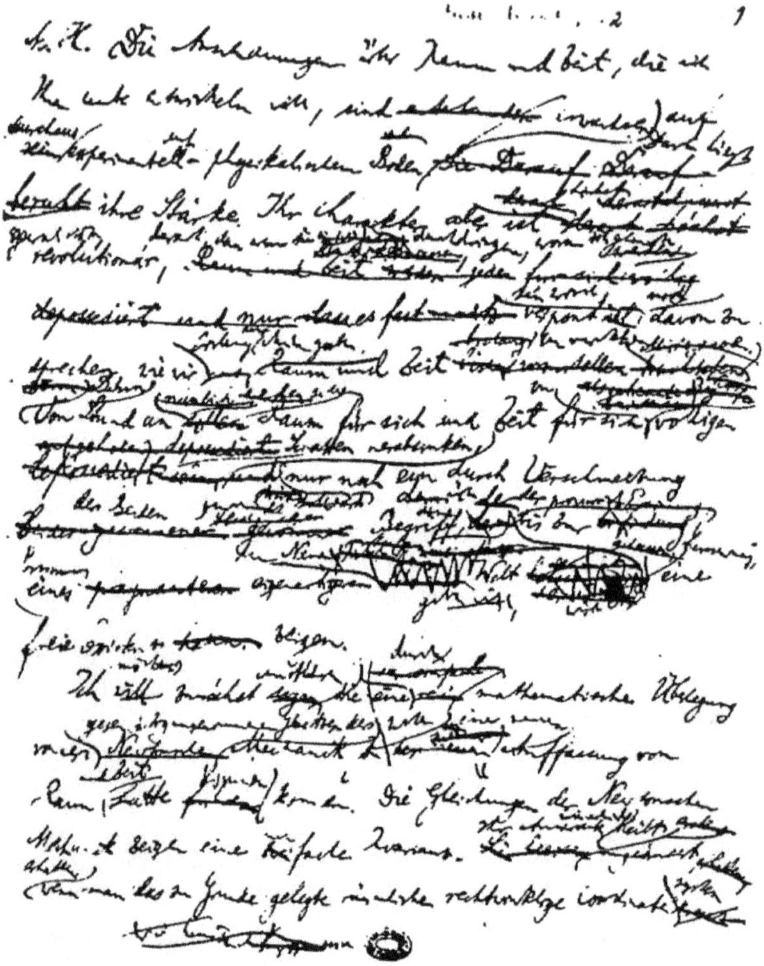

Figure 3: First Page: "Space and Time."
First page of a draft version (Draft RZ 2, p.1, labeled 1) of "Space and Time."

more appropriate to a political tract than a discussion of physical theory, yet in the draft the sentence is even stronger. The character of his new views on space and time, Minkowski writes, "is mightily revolutionary, to such an extent that when they are completely accepted, as I expect they will be, it will be disdained to still speak about the ways in which we have tried to understand space and time."[46] The theme of the opening paragraph to "Space and Time" is thus closely related to that of "The Principle of Relativity." Where "Space and Time" speaks of "a kind of union" of space and time, "The Principle of Relativity" presents the world as a "four-dimensional manifold." Although the wording is different to accommodate the different audiences of the two lectures, the idea is the same: beyond the divisions of time and space which are imposed on our experience, there lies a higher reality, changeless, and independent of observer.

The Pre-Established Harmony between Mathematics and Physics

Minkowski hoped to reformulate the physics of Einstein, Lorentz, and Poincaré to yield a "world" with a "free existence." But clearly it would not be enough for Minkowski to use the fact that he thought geometrically to justify a transition of our concepts of space and time. Instead, he grounds his belief in the truth-revealing power of mathematics (geometry in particular) in what he calls the "pre-established harmony between mathematics and physics." This allows him to isolate and investigate mathematical elements of a theory with the faith that in coming back to physical reality the results will be valid and fruitful.

Unfortunately, Minkowski does not discuss the philosophical origins and implications of the "pre-established harmony"; however, his references to it abound. For instance, in the lectures on complex analysis Minkowski asserts that "there emerges a pre-established harmony among the current mathematical branches of knowledge, that is, the specific conceptual schemes [*Begriffsbildungen*] and problems which have proven themselves valuable for the further development of the theory also prove themselves to be fundamental through the development of physical theories."[47] And in the conclusion to "Space and Time", Minkowski assures the reader that "in the development of its mathematical consequences there will be ample suggestions for experimental verifications of the...[relativity principle], which will suffice to conciliate even those to whom the abandonment of old

[46] "Ihr Character ist höchst gewaltig revolutionär, derart, dass wenn sie durchdringen, woran ich glaube, es verpönt sein wird, noch davon zu sprechen wie wir bislang uns Mühe gaben, Raum und Zeit zu verstehen." Draft RZ 2, p. 1, labeled 1(Figure 3).

[47] "Es tritt darin eine prästabilierte Harmonie der mathematischen Wissenszweige zu Tage, dass die nämlichen Begriffsbildungen und Probleme, welche für die Weiterführung der Theorie sich als wertvoll erweisen, auch durch die Entwicklung der physikalischen Theorien sich als fundamental aufdrängen," *Introductory lecture on complex analysis, op cit,* p. 7.

established views is unpleasant or painful, by the idea of a pre-established harmony between pure mathematics and physics."[48]

References to the pre-established harmony appear throughout the draft versions of "Space and Time" as well. In one draft, Minkowski refers to the "pre-established harmony between pure mathematics and nature."[49] But perhaps the most striking discussion of his belief in the power of mathematics to lead us to an understanding of physical reality appears in another draft of "Space and Time": "Electrical theory seems, like no second branch of physics, to be predisposed...for triumphs of pure mathematics. In the world of pure ether, the most fragile mathematical structures ╟*Bildungen*╢ seem to attain complete ╟reality╢ life, whereas everywhere else (one thinks for example of hydrodynamics) the mathematical formulation and problems prove to be only a distant, idealized approximation to crude reality."[50] For Minkowski the investigation of pure mathematics was fundamentally tied to a search for truth in the physical world, and it is mathematical *Bildungen* - literally pictures - which will reveal nature's secrets.

Minkowski's faith in the pre-established harmony dictated the structure and emphasis of his relativity work. The text of "Space and Time," for example, applies the pre-established harmony immediately after the introduction. "First of all, I should like to show how it might be possible," Minkowski began, "setting out from the accepted mechanics of the present day, along a purely mathematical line of thought, to arrive at changed ideas of space and time."[51] Physicists ignore the geometry implicit in physics, he claims, perhaps because by the time they come to study mechanics, they no longer question the axioms of elementary geometry.[52] Minkowski, however, despite his earlier remark that the new physics rests on the ground of "experimental physics," immediately reveals that his interests are obviously mathematical. In any case, "the equations of Newton's mechanics exhibit a two-fold invariance. Their forms remain unaltered, firstly, if we subject the

---

[48] "Bei der Fortbildung der mathematischen Konsequenzen werden genug Hinweise auf experimentelle Verifikationen des Postulates sich einfinden, um auch diejenigen, denen ein Aufgeben altgewohnter Anschauungen unsympathisch oder schmerzlich ist, durch den Gedanken an eine prästabilierte harmonie zwischen der reinen Mathematik und der Physik auszusöhnen." *op. cit.*, RZ, p. 66; ST, p. 91.

[49] Draft RZ 4, p. 22, labeled 20.

[50] "Die Elektrizitätslehre scheint wie kein zweites Gebiet der Physik prädisponiert...für Triumphe der reinen Mathematik. Während anderwärts, man denke z. B. an die Hydrodynamik, die rnathematischen Formulierungen und Problemstellungen nur als entfernte ideale Annäherungen an die rohe Wirklichkeit sich erweisen, scheinen in der Welt des reinen Äthers die zartesten mathematischen Bildungen vollkommenes ╟Realität╢ Leben zur erlangen." Draft RZ 4, p. 1, labeled 1. ╟Crossed brackets╢ indicate the enclosed word was crossed out in the original.

[51] "Ich möchte zunächst ausführen, wie man von der gegenwärtig angenommenen Mechanik wohl durch eine rein mathematische Überlegung zu veränderten Ideen über Raum und Zeit kommen könnte." *op. cit.*, RZ, p. 54; ST, p.75.

[52] *ibid.*

underlying system of spatial coordinates to any arbitrary *change of position*; secondly, if we change its state of motion, namely, by imparting to it any *uniform translatory motion*; furthermore, the zero point of time is given no part to play." [53] Minkowski then explains that the first group of transformations shows the invariance of the form of the equations of motion under displacements or rotations of the coordinate system. This group's validity is based on two geometric presuppositions about space: homogeneity and isotropy. The invariance of the second group, that of uniform translation, is a physical property based on the assumption that there is no mechanical phenomenon which allows us to distinguish a preferred inertial system. On the one hand, we have a group of geometric transformations; on the other, a group of physical ones: "Thus the two groups, side by side, lead their lives entirely apart. Their utterly heterogeneous character may have discouraged any attempt to compound them. But it is precisely when they are compounded that the complete group, as a whole, gives us to think."[54]

The goal of Minkowski's investigation is to understand the "complete group." To this end, "we want to visualize the relationships graphically." If we let x, y, z be the space coordinates and t be time, we can represent a point of space at a particular time by (x, y, z, t). The object given by a particular value of such a quadruple Minkowski calls a "world-point." The collection of all world-points constitutes the world, and the world-points that trace a single object's existence in space and time are christened its "world-line."

Though an appreciation of four dimensions seems difficult to the uninitiated, Minkowski insists that "with this most valiant piece of chalk I might project upon the blackboard four world-axes. Since merely one chalky axis, as it is, consists of molecules all a-thrill, and moreover is taking part in the earth's travels in the universe, it already affords us ample scope for abstraction; the somewhat greater abstraction associated with the number four is for the mathematician no infliction." [55] The mathematician's four dimensions are no less real for Minkowski than the physicist's three. In fact, the invariance and timelessness of the world-line give Minkowski reason to believe that the four-dimensional manifold is a higher reality than the three

---

[53] *ibid.* "Die Gleichungen der Newtonschen Mechanik zeigen eine zweifache Invarianz. Einmal bleibt ihre Form erhalten, wenn man das zugrunde gelegte räumliche Koordinatensystem einer beliebigen *Lagenveränderug* unterwirft, zweitens, wenn man es in seinem Bewegungszustande verändert, nämlich ihm irgendeine *gleichförmige Translation* aufprägt; auch spielt der Nullpunkt der Zeit keine Rolle."

[54] "So führen jene zwei Gruppen ein völlig getrenntes Dasein nebeneinander. Ihr gänzlich heterogener Charakter mag davon abgeschreckt haben, sie zu komponieren. Aber gerade die komponierte volle Gruppe als Ganzes gibt uns zu denken auf." *op. cit.*, RZ, p. 54; ST, p. 76.

[55] Ich könnte mit kühner Kreide vier Weltachsen auf die Tafel werfen. Schon *eine* gezeichnete Achse besteht aus lauter schwingenden Molekülen und macht zudem die Reise der Erde im All mit, gibt also bereits genug zu abstrahieren auf; die mit der Anzahl 4 verbundene etwas grössere Abstraktion tut dem Mathematiker nicht wehe." *op. cit.*, RZ, p. 55; ST. p. 76.

dimensions we perceive. Then, almost as if to reassure us of the physical reality of the manifold, Minkowski immediately adds, "Not to leave a yawning void anywhere, we will imagine that everywhere and at every time there is something perceptible. To avoid saying 'matter' or 'electricity' I will use for this something the word 'substance'."[56] When Minkowski places a "substance" at every point in space-time, he is looking back to the ponderomotive ether, refusing to discard the concept entirely although it has been voided of qualities and properties.

Out of these considerations came Minkowski's most prophetic comment, which implies that physical law will someday be expressible as laws of world-lines. "The whole universe is seen to resolve itself into similar world-lines, and I would fain anticipate myself by saying that in my opinion physical laws might find their most perfect expression as reciprocal relations between these world-lines."[57] Although Minkowski's prediction did not anticipate the field-theoretical direction general relativity would eventually take, his view has had far-reaching consequences, especially for the theory of interacting particles.

The Role of Aesthetic Criteria

If one grants that Minkowski can pass from good mathematics to productive physics, it remained for him to ground the new physics on mathematics alone. He accomplishes this by comparing Newtonian and relativistic theories on the basis of three criteria of geometrical elegance that emerge from his visual thinking: symmetry, generality, and invariance. Together they seem to form the motivation and the justification for Minkowski's adoption of the new physics.

Symmetry

In "The Principle of Relativity," immediately after the introduction and brief remarks on the Michelson-Morley experiment, Minkowski turns to the question of symmetry. Symmetry plays a complex and vital role in the development of relativity; it is therefore important to distinguish between a variety of concepts which fall under its name. The symmetry Minkowski wishes to point out emerges from the basic equations of the Lorentz theory of the electron, which "possess even a further symmetry (other than that they are independent of any particular orthogonal coordinate system in space). In the

---

[56] "Um nirgends eine gähnencle Leere zu lassen, wollen wir uns vorstellen, dass aller Orten und zu jeder Zeit etwas Wahmehmbares vorhanden ist. Um nicht Materie oder Elektrizität zu sagen, will ich für dieses Etwas das Wort Substanz brauchen." *op. cit.*, RZ, p. 55; ST, p. 76. (Translation slightly modified.)

[57] "Die ganze Welt erscheint aufgelöst in solche Weltlinien, und ich möchte sogleich vorwegnehmen, dass meiner Meinung nach die physikalischen Gesetze ihren vollkommensten Ausdruck als Wechselbeziehungen unter diesen Weltlinien finden dürften." *op. cit.*, RZ, p. 55; ST, p. 76.

usual way these equations are written, this symmetry is not made explicit. From the outset I want here to exhibit this symmetry, which none of the others did, not even Poincaré. In this way, I believe, the form of the equations will become clear."[58]

The new symmetry argument referred to by Minkowski differed from those already employed by Einstein and Poincaré. For Einstein it was objectionable that the same phenomena observed from two different reference frames were given fundamentally different physical explanations. Einstein writes: "It is all the same whether the magnet is moved or the conductor; only the relative motion counts according to the Maxwell-Lorentz theory. However, the theoretical interpretation of the phenomena in these two cases is quite different....The thought that one is dealing here with two fundamentally different cases was for me unbearable [*war mir unerträglich*)."[59] Thus Einstein objected to current theory on the basis of an abhorrence of unnecessary asymmetry and complexity rather than because of any inadequacy of experimental prediction.[60] It is less clear which of Poincaré's symmetries Minkowski has in mind. He could be referring either to Poincaré's proof that the Lorentz transformations form a Lie group, or to the formal symmetry between the space and time variables Poincaré presents in the gravitation section of his 1906 paper.

Minkowski, like Einstein, objected to the prevailing theory on what could be called aesthetic grounds. He objected to a lack of symmetry in the old physics, but a lack of geometric, rather than physical symmetry. Minkowski's new, geometrical symmetry is grounded in Poincare's *x, y, z, ict* formalism. In "The Principle of Relativity", Minkowski begins with Poincaré's four-space and goes on to show that the Lorentz transformation is an orthogonal transformation for all vectors which transform like *x, y, z, t.* Finally he reasons that physical laws composed of these four vectors will be covariant. (It is not clear why he does not finish the task of putting the Maxwell equations in covariant form.) He claims that covariance follows from the Lorentz transformation alone; that is, without any discussion of the status of the relativity principle. As he puts it, covariance follows "as a pure triviality, that is without the introduction of any new, previously unincluded

---

[58] "...besitzen, ausser dass sie natürlich von der Wahl eines rechtwinkligen Koordinatensystems im Raume unabhängig sind, noch eine gewisse weitere Symmetrie, die bei der gewöhnlichen Schreibweise nicht zum Ausdruck gebracht wird. Ich will hier, was übrigens bei keinem der genannten Autoren, selbst nicht bei Poincaré, geschehen ist, jene Symmetrie von vornherein zur Darstellung bringen, wodurch in der Tat die Form der Gleichungen, wie ich meine, äusserst durchsichtig wird," *op. cit.*, pp.928-29.

[59] Einstein, Albert, "Fundamental Ideas and Methods of Relativity Theory. Presented in Their Development," pp. 20-21, cited in Holton, *Thematic Origins, op cit*, pp. 363-64.

[60] For more on symmetry in Einstein's work see Miller, "Einstein's Relativity Paper," *op. cit.*, pp. 11, 14, 23; and Holton, "On Trying to Understand Scientific Genius," *Thematic Origins, op. cit.*, esp. pp. 362-67.

law...".[61] Only in the next section, on matter, does he introduce the "new law" of relativity.

Different observers assign different coordinates to a given event. Minkowski reasons that since $t^2 - x^2 - y^2 - z^2$ is Lorentz-invariant, the four-dimensional hyperboloid,

$$t^2 - x^2 - y^2 - z^2 = constant$$

represents the set of all possible space-time coordinates of one event.[62] The principle of relativity tells us that "absolute rest corresponds to no properties of the phenomena." Since in four dimensions there is a non-zero vector lying on the hyperboloid and corresponding to zero velocity, any point (x, y, z, t) on the hyperboloid can be transformed to lie on the t-axis. Such a Lorentz transformation will take the hyperboloid back into itself. This is the geometric symmetry which Minkowski introduces into relativity. Its physical consequence is that no particular measurement of the coordinates of an event can indicate absolute rest.

Alternatively, Minkowski adds, one should consider the velocity four-vector. Then the new symmetry begins with the fact that the zero three-velocity vector is simply another vector on the hyperboloid. Since this is true for any four-vector, and we can use four-vectors to specify fully the physical characteristics of a system, both electrodynamic and mechanical, we can see that this "further symmetry" is perfectly general.

The four-dimensional representation places rest and motion on equal graphical footing. Since any four-vector can be transformed to the "rest-vector," leaving the hyperboloid of the appropriate invariant unchanged, the principle of relativity, i.e., that no phenomena are attached to absolute rest, stands fully exposed. Such a symmetry is clearly distinct from the physical symmetry of Einstein and the formal or group symmetries of Poincaré. Minkowski's graphical symmetry is not, however, the only geometric consideration he wishes to present.

Generality

In "Space and Time", where the space-time concept is more fully explained, Minkowski directs attention to the added generality of the relativistic transformations when seen graphically, through space-time. To emphasize this element of generality, he begins the paper by making a more detailed study of the structure of space and time in classical physics.

---

[61] RP, *op. cit.*, p. 931.

[62] Note that while Minkowski does not make this clear in the text, he obtained the hyperboloid only for *homogeneous* linear transformations (fixed origins). In the case of velocity- or momentum-space, the problem does not arise since we are dealing with differences of positions. I am indebted to John Stachel for this observation.

Minkowski first points out that the laws of classical physics are invariant under two sets of transformations, one geometric, the other physical. The second group, transformations of inertial reference systems, can be written:

$$x' = x - At$$
$$y' = y - Bt$$
$$z' = z - Ct$$
$$t' = t$$

Minkowski then seems to assume one has in mind a kind of "Galilean space-time diagram," for he immediately adds, "The time axis can, from now on, have a completely free orientation towards the upper half World $t > 0$." Though Minkowski does not graph the Galilean space-time diagram, it is helpful to present it to illustrate this argument (see Figure 4).[63] Figure 4 makes clear that since there is no restriction on velocity in the Galilean transformations, all values of v from plus infinity to minus infinity are possible. By the concept of absolute time all observers must agree on the lines of simultaneity which run parallel to the x-axis. For all systems where $x = x'$, we would then have the situation shown in Figure 5. The Galilean transformation is thus characterized by a completely free t and an unchanged x.

Minkowski intends to show that Galilean space-time can be understood as a special case of a more general geometric structure. To this end he considers the invariant form, $c^2t^2 - x^2 = 1$, which in the x-t plane appears as an equilateral hyperbola. For the moment, Minkowski attaches no physical meaning to x, t, and c, viewing the transformations purely from a formal standpoint. In Figure 6 Minkowski constructs a linearly transformed coordinate system in which the form, $c^2t^2 - x^2 = 1$ is preserved. This may be seen from the construction: in Figure 6 let t', x' be arbitrarily, symmetrically inclined with respect to the "light-line" $ct - x = 0$. Call the t' intersection with the hyperbola, A'. Construct the tangent to the hyperbola at A', and call its intersection with the light-line B'. Then complete the parallelogram and label as shown in Figure 6. Now A' is on the hyperbola, so we demand $c^2t'(A')^2 - x'(A')^2 = 1$, where $t'(A')$ is the t' coordinate of A'. But since A' is on the t'-axis, $x'(A') = 0$, so $c^2t'(A')^2 = 1$, whence $t'(A') = 1/c$. $\overline{A'B'}$ is parallel to x' so $t'(B') = t'(A') = 1/c$ and by definition we know $x' = ct'$ along the light-line for all x, t systems. Therefore, $ct'(B') = x'(B') = 1$. It follows, since $\overline{C'B'}$ is parallel to the t'-axis, that $x'(C') = 1$. In sum t' and x' are uniquely scaled by the demand that the quadratic form $c^2t^2 - x^2 = 1$ in all frames. We thus have a well-defined transformation group with parameter c.

---

[63] I have borrowed the diagram in Figure 4 from Born, Max, *Einstein's Theory of Relativity* (New York: Doyer, 1962), p. 75.

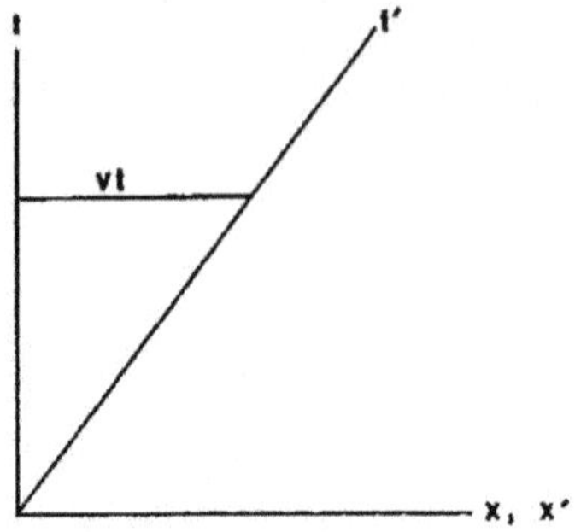

Figure 4: Galilean Space-Time

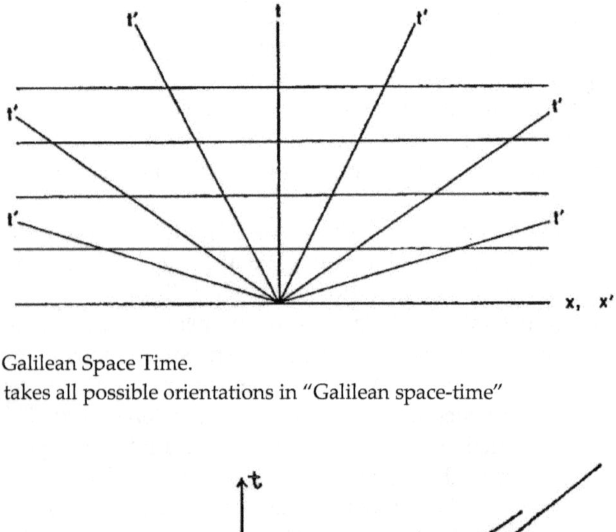

Figure 5: Galilean Space Time.
Time axis takes all possible orientations in "Galilean space-time"

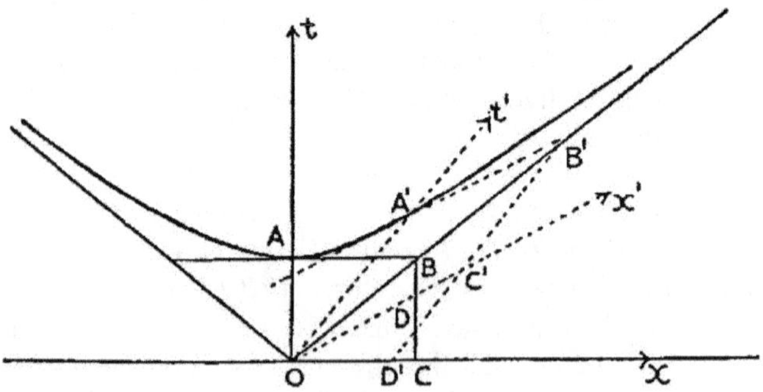

Figure 6: Space-Time Diagram
The diagram illustrates construction of a new time axis from the hyperboloid now known as the "calibration curve." (From "Space and Time": RZ, p.56; ST, p.78.

But all is not mathematical artifice. Minkowski tells us that *c* is the velocity of light, with the same value in all frames of reference, i.e., $ct' - x' = ct - x = 0$ along the light-line. Finally Minkowski notes that an easy calculation shows that $\overline{OC/OC'} = \sqrt{1 - v^2/c^2}$.[64] Thus given a reference system S' and a scale in S we have a graphical method for determining $S''s$ scale. These transformations form a group with parameter *c* which Minkowski calls G_c.

Now suppose $c \to \infty$. Then the hyperbola "calibration curve" degenerates to a straight horizontal line as its tangent $\left(\frac{dt}{dx}\right) = \left(\frac{1}{c}\right)\left(\frac{x}{(1+x^2)^{½}}\right) \to 0$. Minkowski then uses this fact to show where the freedom of the *t*-axis in Newtonian mechanics comes from:

> If we now allow *c* to increase to infinity, and 1/*c* therefore to converge towards zero, we see from the figure that the branch of the hyperbola bends more and more towards the axis of *x*, the angle of the asymptotes becomes more and more obtuse, and that in the limit this special transformation changes into one in which the axis of *t'* may have any upward direction whatever, while *x'* approaches more and more exactly to *x*.[65]

In more explicit form this may be seen as follows: the tangent at *A'* defines the lines of simultaneity since it is parallel to *x'*. ($c^2t'^2 - x'^2 = 1$; $dt'/dx' = 0$ at $x' = 0$, so the tangent there is parallel to the *x'*-axis as claimed.) Thus as *c* goes to infinity, the lines of simultaneity approach the horizontal and coincide for all values of *v*. We are left with the situation portrayed in Figure 5. All frames agree on simultaneity, *x* is left fixed, and is totally free. Gt_∞ is thus the limiting case of G_c corresponding to a degenerate calibration curve - the horizontal straight line. This is the space-time structure of Newton's absolute space and time.

---

[64] Sommerfeld's derivation is given in the appendix to *The Principle of Relativity*, *op. cit.* Using the above considerations we can deduce that $c^2t^2 - x^2$ is an invariant (i.e., has the same value for all observers). On the light line $x = ct$, $x' = ct'$ by the invariance of the speed of light. Then $x' - ct' = 0$, so that by the Lorenz transformations (where $\gamma = \frac{1}{\sqrt{1-\left(\frac{v}{c^2}\right)}}$, $x' - ct' = \gamma \cdot (x - vt) - \gamma c\left(t - \frac{v}{c^2 x}\right) = \gamma \cdot \left[x\left(1 + \frac{v}{c}\right) - ct\left(1 + \frac{v}{c}\right)\right] = \gamma \cdot \left(1 + \frac{v}{c}\right)(x - ct)$. Conversely, if $x = -ct$ and $x' = -ct'$, reverse $v/c \to -v/c$ but $\gamma(v/c) = \gamma(-v/c)$, so $x' + ct' = \gamma \cdot \left(1 - \frac{v}{c}\right)(x + ct)$. Multiplying the two expressions yields $x'^2 - c^2t'^2 = -\gamma'^2 \cdot \left(1 - \frac{v}{c^2}\right)(x^2 - c^2t^2) = x^2 - c^2t^2$ since $\gamma'(1 - v/c^2) = 1\ Q.E.D.$

[65] "Lassen wir jetzt *c* ins Unendliche waschen, also 1/*c* nach Null kovergieren, so leuchtet an der beschreibenen Figur ein, dass der Hyperbelast sich immer mehr der *x*-asche anschmiegt, der Asymptotenwinkel sich zu einen gestreckten verbreitert, jene spezielle Transformation in der Grenze sich in eine solche verwandelt, wobei die *t'*-Asche eine beliebige Richting nach oben haben kann und *x'* immer genauer sich an der *x* annähert." *op. cit.*, RZ, p.56; ST, p.78.

From his belief in the "pre-established harmony" and his discovery of these geometrically satisfying properties, Minkowski concludes that the four-dimensional theory is superior to Newtonian three-dimensional physics. In fact, after discussing generality and symmetry, he claims: "This being so, and since G_c is mathematically more intelligible than G_∞, it looks as though the thought might have struck some mathematician, fancy-free, that after all, as a matter of fact, natural phenomena do not possess an invariance with the group G_∞, but rather with a group G_c, c being finite and determinate, but in ordinary units of measure, *extremely great*."[66] Strictly speaking, the Galilean transformations are perfectly well defined. It is therefore quite significant that Minkowski considers the group G_c "mathematically more intelligible" (*verständlicher*); he seems to be referring to the added generality and symmetry of the geometric interpretation and not to the mathematical consistency of the transformation.

In what probably is a draft of the introduction to the "*Grundgleichungen*", we find the same sentiments about the new transformation law expressed somewhat differently: "As a result of the progress which pure mathematics has made in the last century, it is particularly easy for the pure mathematician, easier than for the modem physicist, to assimilate the new law and to be enthusiastic about it. This is because it is in fact ‖⊢mathematically⊣‖ theoretically in many respects more satisfying than the Galilean law."[67] Once again Minkowski stresses the familiarity of the mathematician with non-Euclidean geometries and quadratic transformations in responding to the mathematical content of the new physics. Instead of describing the new representation as "mathematically more intelligible" (*mathematisch verständlicher*), he now says "theoretically more satisfying" (*theoretisch befriedigender*), an even stronger claim. This stronger claim contrasts especially with the earlier word choice in the manuscript modestly describing the new representation as "mathematically more satisfying."

Minkowski implies that considerations of symmetry and completeness could have suggested to a mathematician that G_c rather than G_∞ is the correct transformation group for the *physical* world: "Such a

---

[66] "Bei dieser Sachlage, und da G_c mathematisch verständlicher ist als G_∞, hätte wohl ein Mathematiker in freier Phantasie auf den Gedanken verfallen können, dass am Ende die Naturerscheinungen tatsächlich eine Invarianz nicht bei der Gruppe G_∞, sondern vielmehr bei einer Gruppe G_c mit bestimmtem endlichen, nur in den gewöhnlichen Masseinheiten *äussert grossen c* besitzen." *op. cit.*, RZ, p.56, ST, p. 79.

[67] "Bei der Entwickelung, welche die reine Mathematik im letzten Jahrhundert genommen hat, fällt es dem reinen Mathematiker besonders leicht, leichter als dem modernen Physiker, sich in das neue Gesetz hineinzudenken und sich darür zu enthusiasmieren, weil es in der Tat ‖⊢mathematically⊣‖ theoretisch sich in vielen Hinsichten viel befriedigender anlässt als das Galileische Gesetz." Draft "Grundgleichungen," p. 2. See appendix "Notes on Manuscript Sources."

premonition would have been an extraordinary triumph for pure mathematics. Well, mathematics, though it now can display only staircase-wit, has the satisfaction of being wise after the event, and is able, thanks to its happy antecedents, with its senses sharpened by an unhampered outlook to far horizons, to grasp forthwith the far-reaching consequences of such a metamorphosis of our concept of nature."[68] By a kind of "after-wit," mathematics broaches the subject only after physical considerations have led to the transformations. But now that such foundations have been laid, mathematics, thanks to the development of space-time, can be used to explore the implications of the transformation of our views of nature. Just as in "The Principle of Relativity," mathematics is essential here because it serves to present our conception of nature (*Naturauffassung*) in a simpler, more coherent fashion than physical concepts with direct empirical content. One suspects that in Minkowski's view, pure thought in the form of mathematics, is capable of advancing our concept of reality.

The Invariant

Among the three aesthetic criteria employed by Minkowski in his physics - symmetry, generality, and invariance - invariance is the most important aspect of relativity theory brought out by the space-time formulation. There is evidence that concern for invariance, along with visualization, was noticed by his contemporaries as a salient feature of Minkowski's thinking.[69] Minkowski himself began to explore the conceptual simplifications afforded by the new formulation in his first relativity paper, "The Principle of

---

[68] "Eine solche Ahnung wäre ein ausserordentlicher Triumph der reinen Mathematik gewesen. Nun, da die Mathematik hier nur mehr Treppenwitz bekundet, bleibt ihr doch die Genugthuung, dass sie dank ihren glücklichen Antezedenzien mit ihren in freier Fernsicht geschärften Sinnen die tiefgreifenden Konsequenzen einer solchen Ummodelung unserer Naturauffassung auf der Stelle zu erfassen vermag." *op. cit.*, RZ, p. 57; ST, p. 79.

[69] This evidence emerges from an unlikely but amusing source - a student parody of the Göttingen course catalogue written around 1907. The first reference mocks Minkowski's application of geometry to number theory by having him present a "Chemical Number Theory": "H. Minkowski: Chemical Number Theory (self-advertisement). I can no longer hold back from the mathematical world one of the most interesting results of my application of number theory to chemistry. It concerns the 'periodic system' of the elements which, as everybody knows, is visualized through the following curve....[Minkowski graphs atomic volume against atomic weight.] The result becomes clear through the latest surprising results of Hilbert...and draws on the function I introduced earlier: ?(x), !(x), ;(x), = (x) which follows from ...

$$\int_{z(a)}^{z(e)} log \sqrt{?\,(y)^{(z)}} dz.$$

My detailed textbook about these matters should appear in the course of the century." The second page parodies Minkowski's insistence on the invariant: "On the invariants of the Göttingen shooting match with special attention to the *Moppenonkels,* 6 hours, private lessons to be arranged." Maybe the invariants of the Göttingen shooting match are just what particle physics has been waiting for. I would like to thank Mrs. L. Rüdenberg for making a copy of this document available to me.

Relativity." But in "Space and Time", a new understanding of the mathematical structure of relativity leads Minkowski much further, to a specific world view. This metaphysical standpoint, "the theory of the Absolute World" as he calls it, involves two physical concepts: covariance and invariance.

An equation is covariant if the form of the equation remains the same in all inertial reference systems, that is, if its variables transform like *x, y, z, t.* An expression is invariant if it equals a scalar as defined in vector analysis, for example, $c^2t^2 - x^2 = c^2t'^2 - x'^2$ (as shown earlier in footnote 64).[70] Some confusion may arise since Minkowski uses the word "*Invarianz*" to refer to both covariance and invariance, but it is always clear from the context which of the two he means. Here we will use modern terminology to avoid any ambiguity.

The existence of invariants for the relativistic transformation forms the third aesthetic criterion Minkowski considers in his four-dimensional relativistic theory. "The innermost harmony of these [electrodynamic] equations," he writes, "is their invariance under the transformations of the expression $dx^2 + dy^2 + dz^2 - dt^2$ into itself."[71] In Newtonian space-time the free *t*-axis prevents us from constructing such an invariant expression. Like symmetry and generality, invariance is an aesthetic geometric criterion which supports the new conception of space-time. Minkowski's belief in the pre-established harmony allows him to focus his attention on the mathematics that underlies relativity. Then, Minkowski claims, by applying criteria such as symmetry, generality, and invariance to the mathematics, we can discover essential elements of our *physical* universe.

The Theory of the Absolute World

Minkowski's success in translating the laws of physics into space-time led him to believe that the new formulation of physics demanded a revision of our metaphysical views as well. Minkowski endowed abstract space-time with the reality previously accorded three-dimensional space and called the result "The Theory of the Absolute World." Minkowski saw the Absolute World as so important that he wondered what had prevented other physicists from discovering it. As a partial answer, Minkowski suggested that Einstein and others had criticized space and time in isolation rather than as parts of a whole.

Certainly Minkowski acknowledged Einstein's role in demonstrating that "proper time" is more than a mathematical device. Through this demonstration "time, as a concept unequivocally determined by

---

[70] Eisenhart, Luther Pfahler, *Riemannian Geometry* (Princeton: Princeton University Press, 1966), p. 6. Not to be confused with an algebraic invariant.

[71] "Die innerste Harmonie dieser Gleichungen aber ist ihre Invarianz bei den Transformationen des Ausdrucks $dx^2 + dy^2 + dz^2 - dt^2$ in sich." Draft RZ 4, p. 17, labeled 26.

phenomena, was first deposed from its high seat."[72] Under relativity, Minkowski asserts, "time" as a concept independent of reference frame has no meaning. We are left with "times" instead of "time." Nevertheless this vital contribution of Einstein's did not take us far enough, for "neither Einstein nor Lorentz made any attack on the concept of space, perhaps because in the above-mentioned special transformation, where the plane of *x', t'* coincides with the plane of *x, t*, an interpretation is possible by saying that the *x*-axis of space maintains its position."[73] To explain why no one, including Einstein, had attacked the whole concept of "space," Minkowski conjectures the following: in one dimension a relativistically correct solution can be obtained by leaving *x* and *x'* superimposed and rotating *t'* through the appropriate angle (Figure 7a). Because they concentrated on this special one-dimensional solution, Minkowski speculates, previous authors neglected the structure of *space*, his own central concern.

Minkowski sees his criticism of space as the logical complement of Einstein's criticism of time. Where Einstein granted reality to each "time" of an observer, Minkowski gives each observer's "space" a similar reality. "We should then have in the world no longer *space*, but an infinite number of

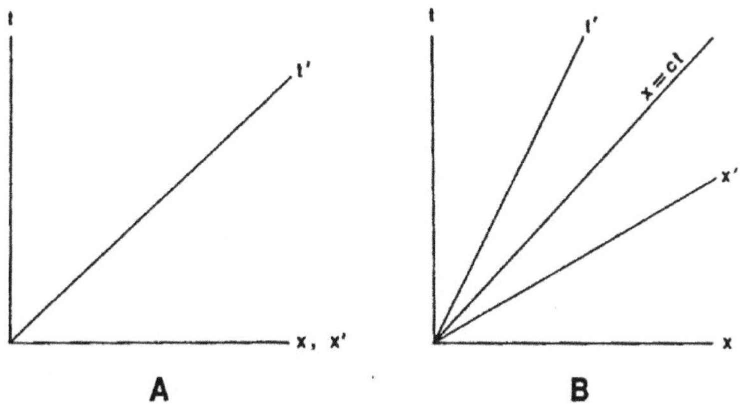

Figure 7: Einstein and Space.
The possible superimposition of *x* and *x'* axes might account for Einstein's attacking time and not space (7a). The "usual" Minkowski space-time diagram (7b).

[72] "Damit war nun zunächst die Zeit als ein durch die Erscheinungen eindeutig festgelegter Begriff abgesetzt." *op. cit.*, RZ, p. 60; ST, pp. 82-83.

[73] " An dem Begriffe des Raumes rüttelten weder Einstein noch Lorentz, vielleicht deshalb nicht, weil bei der genannten speziellen Transformation, wo die *x', t'*-Ebene sich mit der *x, t*-Ebene deckt, eine Deutung möglich ist, als sei die *x*-Achse des Raumes in ihrer Lage erhalten geblieben." *op. cit.*, RZ, p. 60; ST, p. 83.

spaces, analogously as there are in three-dimensional space an infinite number of planes. Three-dimensional geometry becomes a chapter in four-dimensional physics. Now you know why I said at the outset that space and time are to fade away into shadows, and only a world in itself will subsist."[74] Seen from the standpoint of the four-dimensional world, the particular space-time coordinate system attached to an inertial reference system is a sub-space. If one considers changing values of space and time, "relativity" may be an appropriate name. For although it seems reasonable to speak of different "times," "one may expect to find a corresponding violation of the concept of space appraised as another act of audacity on the part of mathematical culture. Nevertheless, this further step is indispensable for the true understanding of the group G_c and when it has been taken, the word *relativity-postulate* for the requirement of an invariance with the group G_c seems to me very feeble."[75] "Relativity" in Minkowski's view, is relevant only to the particular spaces embedded in the manifold.

Just as the myriad of cross-sections of three-space can be freed from perspective variations only by considering the space as a whole, so three-dimensional physics can only be fully understood in four-dimensional space-time. Thus, when several pages later in "Space and Time", Minkowski compares the four-dimensional representation of electromagnetic force with previous, three-dimensional formulations of the same idea, he notes that "we are compelled to admit that it is only in four dimensions that the relations here taken under consideration reveal their inner being in full simplicity, and that on a three-dimensional space forced upon us *a priori* they cast only a very complicated projection."[76] Minkowski recognizes the difficulty in accepting such a conception of space, but claims that space-time is no less real for being the product of a "mathematical culture." "Since the postulate comes to mean that only the four-dimensional world in space and time is

---

[74] "Hiernach würden wir dann in der Welt nicht mehr *den* Raum, sondern unendlich viele Räume haben, analog wie es im dreidimensionalen Raume unendlich viele Ebenen gibt. Die dreidimensionale Geometrie wird ein Kapitel der vierdimensionalen Physik. Sie erkennen, weshalb ich am Eingange sagte, Raum und Zeit sollen zu Schatten herabsinken und nur eine Welt asich bestehen." *op. cit.* (note 25): RZ, p. 57; ST, pp. 79-80.

[75] "Über den Begriff des Raumes in entsprechender Weise hinwegzuschreiten, ist auch wohl nur als Verwegenheit mathematischer Kultur einzutaxieren. Nach diesem zum wahren Verständnis Gruppe G_c jedoch unerlässlichen weiteren Schritt aber scheint mir das Wort *Relativitätspostulat* für die Forderung einer Invarianz bei der Gruppe G_c sehr matt." *op. cit.*, RZ, p. 60; ST, p. 83. I have translated *mathematische Kultur* as "mathematical culture," not, with ST, as "higher mathematics."

[76] "...so wird man nicht umhin können zuzugeben, dass die hier in Betracht kommenden Verhältnisse ihr inneres Wesen voller Einfachheit erst in vier Dimensionen enthüllen, auf einen von vornherein aufgezwungenen dreidimensionalen Raum aber nur eine sehr verwickelte Projektion werfen." *op. cit.*, RZ, pp. 65-66; ST, p. 90. Minkowski cites, as examples of this previous formulation, K. Schwarzschild, *Göttinger Nachrichten* (1903), p. 132, and H. A. Lorentz, *Enzyklopädie der mathematischen Wissenschaft*, 5, Art. 14, p. 199.

given by phenomena, but that the projection in space and in time may still be undertaken with a certain degree of freedom, I prefer to call it the *postulate of the absolute world* (or briefly, the world-postulate)."[77]

When Minkowski completed "The Principle of Relativity" in 1907, he was already intrigued by the invariant. Nonetheless, the elements of the "Absolute World" discussed above in relation to "Space and Time" are not present in the published version of "The Principle of Relativity." The manuscript drafts, by contrast, contain handwritten musings on a name for the four-dimensional hyperboloid: World Surface (*Weltfläche*), World Mirror (*Weltspiegel*), and Cosmograph (*Kosmograph*) (see Figure 8).[78] Each of these names stresses both the universal (*West, Kosmo*) and the visual *(-fläche, -spiegel, -graph)* aspect of Minkowski's theory. Already in "The Principle of Relativity," then, Minkowski was searching for a name which would convey the importance of his discovery, an importance which he felt went beyond a simple rewriting of physical formalisms.

The invariants of the physical theory suggested to Minkowski that there was a world independent of the observer, a *Welt an sich*. In the published papers there are few hints of any psychological connection between the loss of the Newtonian absolutes of space and time and the creation in "Space and Time" of the "Postulate of the Absolute World." But several times in the manuscript versions of the paper, the two "absolutes" are unambiguously linked. In one draft, after introducing his "Principle of the Absolute World," Minkowski writes, "I hope to make plausible [this] essential point: the relations which are connected to the moving point charge become very clear as one abandons the concept [*Vorstellung*] of an absolute time, and passes over to a concept [*Vorstellung*] of an Absolute World as I have explained it."[79] In another draft, he asserts that "there emerges only an Absolute World, but not an absolute space and an absolute time."[80] The shift from the absolutes of space and time, to an Absolute World, also left unchanged another common feature - it maintained a reality not contingent on immediate sense data. Physics has thus gone from the stage of Newton's inaccessible absolutes of mathematical space and time, through the purely

---

[77] "Indem der Sinn des Postulats wird, dass durch die Erscheinungen nur die in Raum und Zeit vierdimensionale Welt gegeben ist, aber die Projektion in Raum und in Zeit noch mit einer gewissen Freiheit vorgenommen werden kann, möchte ich dieser Behauptung eher den Namen *Postulat der absoluten Welt* (oder kurz Weltpostulat) geben." *op. cit.*, RZ, p. 60; ST, p. 83.

[78] Draft RP A, p. 7.

[79] "Ich hoffe...den wesentlichen Punkt plausibel zu machen, dass die Verhältnisse, die mit bewegten punktförmigen Ladungen verbunden sind, sich sehr klären, indem man die Vorstellung einer absoluten Zeit fallen lässt und zu der Vorstellung einer absoluten Welt, wie ich es expliziert habe, übergeht." Draft RZ 4, p. 16, labeled 15.

[80] Es ersteht nur eine absolute Welt, aber nicht ein absoluter Raum und eine absolute Zeit....". Draft RZ 3, p. 5, unlabeled.

formal local time and length contractions of Poincaré and Lorentz, through Einstein's physical interpretation, to Minkowski's Absolute World.

For Minkowski, then, the "Principle of the Absolute World" represented something more than revised notions of space and time. This is made clear in a draft of "Space and Time," where Minkowski writes about his "World Postulate."

> In truth, we are dealing with more than merely a new conception of space and time. The claim is that it is rather a quite specific natural law, which, because of its importance - since it alone deals with the primitive concepts of all Natural Knowledge, namely space and time - can claim to be called the first of all laws of nature. This is a law for which (and I shall explain further) I will coin the expression the "Principle of the Absolute World."[81]

Where Poincaré saw the physics of four dimensions as leading only to a new and "less simple" idiom, Minkowski held it to yield the "first of all laws of nature."

Conclusion

This case study attempts to shed light on the problem of the scientific imagination. The development of space-time shows once again that "crucial experiments" are only partially responsible for precipitating scientific change. In the particular case of Minkowski, the visualization of nature's laws through geometry enters as the primary motivation for the creation of a new physical and metaphysical outlook.

The results of this analysis of the role that aesthetic criteria play in the formation of a scientific theory in many ways parallel Gerald Holton's conclusions concerning Einstein. In his essay "On Trying to Understand Scientific Genius," Holton considers the young Einstein in an attempt to gain insight both into Einstein's mode of thought and into the type of theory he developed.[82] He finds that Einstein employed a visual rather than a verbal type of thinking, which "animates the consideration of symmetries and a corresponding distaste for extraneous complexities from the beginning to the end...".[83]

---

[81] "Es handelt sich genauer gesagt um mehr als bloss eine neue Auffassung von Raum und Zeit vierdimensionale Welt gegeben ist, aber die Projektion in Raum und in seiner Bedeutung nach, weil es allein von den ursprünglichen Begriffen aller Naturerkenntnisse, von Raum und Zeit handelt, den Anspruch erheben darf, das erste aller Naturgesetze genannt werde, ein Gesetz, für welches ich, der ich es Ihnen noch darlege, ein Schlagwort prägen will, indem ich es das Prinzip der absoluten Welt nennen will." Draft RZ 1, p. 1, labeled 2.

[82] Holton, *Thematic Origins, op cit.*, pp. 353-80.

[83] *ibid*, p. 370.

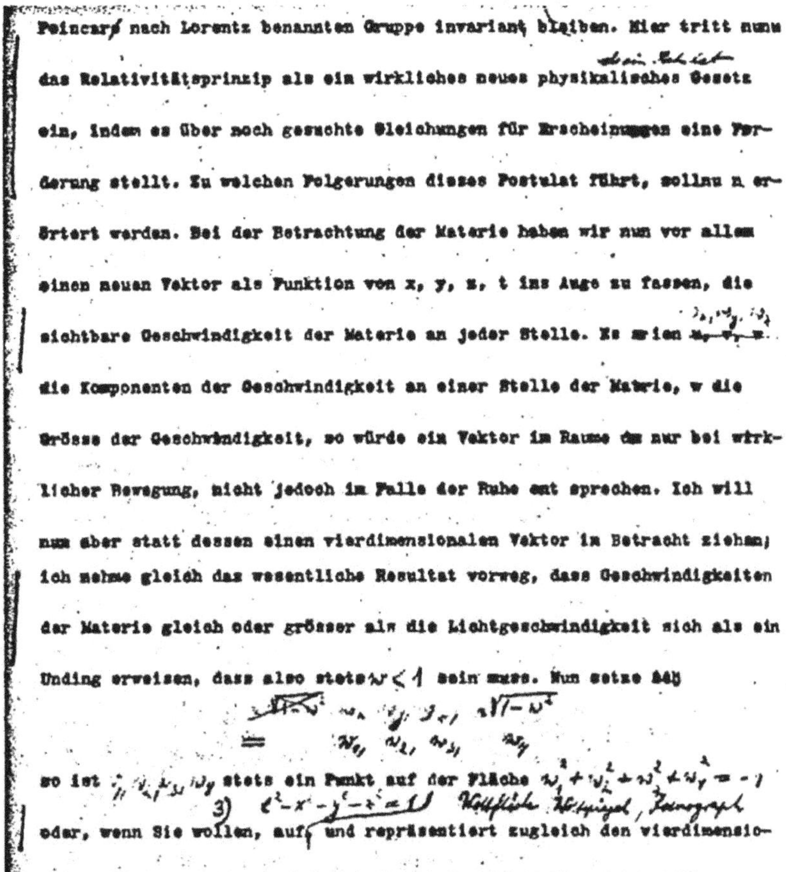
Poincaré nach Lorentz benannten Gruppe invariant bleiben. Hier tritt nun
das Relativitätsprinzip als ein wirkliches neues physikalisches Gesetz
ein, indem es über noch gesuchte Gleichungen für Erscheinungen eine For-
derung stellt. Zu welchen Folgerungen dieses Postulat führt, sollen n er-
örtert werden. Bei der Betrachtung der Materie haben wir nun vor allem
einen neuen Vektor als Funktion von x, y, z, t ins Auge zu fassen, die
sichtbare Geschwindigkeit der Materie an jeder Stelle. Es seien
die Komponenten der Geschwindigkeit an einer Stelle der Materie, w die
Grösse der Geschwindigkeit, so würde ein Vektor im Raume dem nur bei wirk-
licher Bewegung, nicht jedoch im Falle der Ruhe ent sprechen. Ich will
nun aber statt dessen einen vierdimensionalen Vektor in Betracht ziehen;
ich nehme gleich das wesentliche Resultat vorweg, dass Geschwindigkeiten
der Materie gleich oder grösser als die Lichtgeschwindigkeit sich als ein
Unding erweisen, dass also stets $w < 1$ sein muss. Nun setze ich
$$\sqrt{1-w^2}$$
so ist stets ein Punkt auf der Fläche $w_1^2 + w_2^2 + w_3^2 + w_4^2 = -1$
3) $t^2 - x^2 - y^2 - z^2 = 1$ Weltfläche, Weltspiegel, Kosmograph
oder, wenn Sie wollen, auf und repräsentiert zugleich den vierdimensio-
nalen Vektor vom Nullpunkt nach diesem Punkte; und es entspricht auch
der Geschwindigkeit Null, der Ruhe, ein wirklicher derartiger Vektor.
Die Nichteuklidische Geometrie, von der ich schon unbestimmt sprach, ent-
wickelt sich nun für diese Geschwindigkeitsvektoren. Die Transformationen,
von welchen ich vorhin sprach, werden die reellen Transformationen dieses
vierdimensionalen Hyperboloids 3) auf konjugierte Durchmesser, und da ist

Figure 8: World Mirror.
Weltfläche, Weltspiegel, Kosmograph (world. surface, world mirror, and cosmography) - musings on a name for the four-dimensional universe (Draft RP A, p. 7).

In turn, these symmetries play an integral role in both motivating and justifying the transition from Lorentz's theory to Einstein's special relativity, and later from the special to the general theory. Such case studies form a necessary step towards our understanding of change in modem physical theory.

Minkowski, like Einstein, was a visual thinker, though in a different sense. If Einstein may be said to have thought in "concrete" visual terms, running thought experiments through his mind, Minkowski thought in "geometric" visual terms. Where Einstein manipulated clocks, rods, light beams, and trains; Minkowski played with grids, surfaces, curves, and projections.

Characteristic of Minkowski's approach to scientific problems, both mathematical and physical, is his visual-geometric *Anschauung*. The evidence for this characteristic is drawn directly from his work. Minkowski refers to *geometrische Anschauung* in his explicit methodological statements, e.g. in the introductory lecture to the geometry of numbers. The language he employs emphasizes visualization in mathematics: he uses *Begriff* or *Idee* for the physicists' concepts of space and time, while reserving specifically visual words like *Bildung, Begriffsbildung*, and *Vorstellung* for the new concept of space-time. There is the visual imagery of shadows, projections, and planes which Minkowski transfers from three dimensions to the new, four-dimensional space. Finally, the arguments from the geometric elegance of symmetry, generality, and invariance, which Minkowski uses to justify the adoption of space-time, are borrowed from his geometrical background.

Coupled to Minkowski's geometric way of thinking is his belief in a "pre-established harmony" between mathematics and nature which allows him to use his *geometrische Anschauung* to discover the reflection of geometric truths in physics. For Minkowski the role of geometric "aesthetic" criteria is to serve as a hallmark of truth.

Minkowski believed that through thought we can penetrate beyond the appearances of three dimensions to a higher reality; one recalls his image of space and time "sinking into the shadows." We now understand these shadows to be the complicated "projections" of the physical laws whose "inner being" is fully revealed only in the "Absolute World." Such imagery might well be familiar to someone like Minkowski, who received a strict humanistic education in the gymnasium of Königsberg:

> Picture men dwelling in a sort of subterranean cavern with a long entrance open to the light on its entire width. Conceive them as having their legs and necks fettered from childhood, so that they remain in the same spot, able to look forward only, and prevented by the fetters from turning their heads. Picture further the light from a fire burning higher up and at a distance behind them ....

> A strange image you speak of, he said, and strange prisoners.
>
> Like to us, I said. For, to begin with, tell me do you think that these men would have seen anything of themselves or of one another except the shadows cast from the fire on the wall of the cave that fronted them?[84]

But through thought, the philosopher assures us, the prisoners will achieve "the release from bonds, I said, and the conversion from the shadows to the images that cast them and to the light and the ascent from the subterranean cavern to the world above...".[85] Similarly, through the mind's creative eye, through *Anschauung*, Minkowski believed that we can come to know a world free of perspectivism, a four-dimensional world complete, symmetric, and unchanging, that conveys a reality not accessible to us by the direct use of the ever-changing, three-dimensional projection.

Acknowledgements
I am very grateful to Gerald Holton for advice and criticism of this essay, which began in his seminar on case studies in modem physics. At the same time, I would like to acknowledge Erwin Hiebert, Arthur Miller, Joan Richards, John Stachel, James Radner, R. Steven Turner, Gerd Buchdahl, and the referee for *Historical Studies in the Physical Sciences* for all their valuable suggestions. I would like to thank Mrs. Lily Rüdenberg, who was kind enough to speak with me about her father. Her suggestions about finding archival material proved very helpful. I am indebted as well to the Handschriftabteilung of the Niedersächsische Staats- und Universitätsbibliothek Göttingen, especially to Dr. Haenel, Frau Dr. Ritter, and Frau Zirkel for their assistance. At the Niels Bohr Library for the History of Physics, I was greatly aided by Joan Warnow and Peter Dews. I would like to thank Jaklin Kornfilt and Helen Dukas for help with translations. I am grateful to the *Deutsche Akademische Austauschdienst* and the National Science Foundation for their scholarships.

Appendix Notes on Manuscript Sources
In Göttingen, I found several drafts of Minkowski's works on space and time not previously discussed in the literature. They include a set of notes to a seminar conducted around 1905 by Minkowski and Hilbert to review critically the various electron theories.

---

[84] Plato, *Republic,* VII, 514a, *The Collected Dialogues of Plato,* eds., Edith Hamilton and Huntington Cairns (Princeton: Princeton University Press, 1971).
[85] *ibid,* 532b.

Materialist Theories of Time

Richard T.W. Arthur

Historically, materialist thought on time has divided into two opposing trends: those following Democritus, including the Stoics and Gassendi, who see time as that in which matter exists, a kind of precondition for material existence; and those following Aristotle and Epicurus, who see time as an accident of motion or change. Arthur argues that the latter type of theory is to be preferred, being in accordance with the kind of nominalism which he sees as the distinctive characteristic of materialist thought: entities mistakenly thought to exist independently of us and our "temporal" activity -whether souls (la Mettrie), religion-based moral law (d'Holbach), fixed biological species (Diderot, Darwin), gods (Feuerbach), or commodities (Marx)- are revealed to be artifacts resulting from interactions among material things, organisms, or people. Against this background Arthur considers in the third part of the paper the new initiatives in understanding time that have been launched by researchers of quantum gravity, where Barbour and Rovelli argue for the "elimination of time", while identifying Leibniz (together with Mach) as their chief inspiration.

Tempus item per se non est, sed rebus ab ipsis consequitur sensus
Titus Lucretius Carus

Historical Views

What would constitute a materialist theory of time? A first thought might be that, if matter is what exists in space and time, then time should be an existing something within which matter can undergo its changes. According to Aristotle, some of his predecessors thought of time as a container in a quite concrete sense, and others identified it with the motion of the heavens.[1] Thus Pherecydes of Syrus (a contemporary of Anaximander), whom Aristotle regarded as a "hybrid" thinker who wrote only partly in a mythical vein, began his book: "Zas [i.e. Zeus] and *Chronos* [Time] always existed, and so did *Chthonie*; and *Chthonie* acquired the name Earth when Zas gave her the Earth as a bridal gift."[2] (Over)interpreting, perhaps, matter existed even before Zas/Zeus gave it form, and it existed in time, which thus also existed. Of course, it is difficult to know just how to interpret such early texts written in a quasi-mythical vein, and one cannot simply take the personifications to signal literal belief in them as entities. Anaximander, who was one of the originators of prose writing in philosophy, nevertheless regarded time as a

---

[1] Aristotle, *Physics* 218a 33-218b 8. In this paper the following edition will be used: Aristotle, *Physics*, trans. Robin Waterfield (Oxford: Oxford University Press, 1996). See also, Barnes, Jonathan, *The Complete Works of Aristotle*, 2 vols. (Princeton and Oxford: Princeton University Press, 1984), p. 370.
[2] Barnes, Jonathan, *Early Greek Philosophy* (London: Penguin, 1987), p. 58.

kind of arbitrator adjudicating the claims of competing tendencies,[3] whereas Pythagoras is reputed to have called it the soul or procreative element in the universe.[4] Presumably, neither of these personifications of time are supposed to be taken literally, but they are consistent with a worldview in which time (*Chronos*), being an uncreated principle, was regarded as divine.

Nevertheless, there are other indications that these early thinkers thought of time as something quite concrete. Ancient commentators described Pherecydes' *Chronos*, for instance, as a kind of receptacle: it is that in which the things that *Zas* (the creative element) has created out of *Chthonie* (the matter) exist.[5] Again, one surviving line of Empedocles appears to refer to time as the revolving heavenly sphere: "In turn they come to power as time revolves".[6] Similarly, Plato (like Empedocles, much influenced by Pythagorean thought) explicitly identified each planetary orbit as a time, and called time itself a "moving image of eternity".[7] According to these latter formulations, then, time seems to be a concrete thing or motion: either the revolving heaven containing all, or the revolutions of the planets: the month, the year, and the annual cycle of the various planets.

Antiphon was perhaps the first to emancipate himself from this conception of time as a *concretum*. In one of the surviving fragments of his thought, he says: "Time is a thought or measure, not a substance."[8] This notion is in keeping with the general tendency of the thinkers Plato dismissed as "Sophists", to see things which their predecessors had thought of as "by nature" to be in reality "by convention". By the latter half of the fifth century, when Antiphon was active, the conventional element of time measures would in any case have been made more obvious by the various problems with the calendar: Meton had introduced his calendar reform in 432 B.C..[9] At any rate, the idea of time as a measure is taken up by Aristotle, who followed Antiphon in making time measurable, and dependent on a soul to count its measures: "Time is that feature of change which makes number applicable to

[3] Anaximander: "It is necessary that things should pass away into that out of which they came into being. For things must pay one another penalty and compensation for their injustice according to the ordinance of Time." From: Simplicius, *Commentary on the Physics,* §24. from Barnes, *Early Greek Philosophy, op cit,* p. 75.

[4] Whitrow, G. J., *Time in History* (Oxford, New York, Toronto: Oxford University Press, 1988), p. 39.

[5] Freeman, Kathleen, *Ancilla to the Presocratic Philosophers* (Cambridge: Harvard University Press, 1971), p. 39.

[6] Empedocles [B 17.39], Simplicius, *Commentary on the Physics* 1183-4; in Barnes, *Early Greek Philosophy, op cit,* p. 177. Also perhaps suggesting time as a container is "As they [Love and Hate] were formerly, so also will they be, and never, I think, shall boundless Time be emptied of these two." (DK 16) in Freeman, *op cit,* p. 53.

[7] In Plato's *Timaeus,* quoted in Whitrow, *op cit,* p. 19.

[8] (DK 9) in Freeman *op cit,* p.145.

[9] Whitrow, *op cit,* p. 32, 45, 189.

it....[it] is the number of change in respect of before and after."[10] Further in the *Physics*, thus, "if nothing else except mind is such that it can number, it is impossible for there to be time if there is no mind." [11] Aristotle's conceptualism about time is, however, qualified: although time is not change, it is "an aspect of change."[12] Therefore, "it might be possible for there to be change without mind, since before and after are in change, and time is what is before and after insofar as they are numerable." [13] He dismissed the conception of time held by Pherecydes and Empedocles - the sphere in which everything happens - as a reasoning too naïve to be worthy of challenge, and went on to reject Plato's identification of time with process, arguing that whereas movement or change is confined to the particular changing thing and the place it is in, "time is present equally everywhere and with all things".[14] It seems likely that his depiction of time as an aspect of change is the source of Epicurus' later theory of time as an accident of motion.

Now, even though I have suggested that there is evidence among Aristotle's ancient Greek predecessors for what one might call a naïve materialist theory of time, where time is regarded as a concrete thing - a material container, perhaps, or a kind of heavenly clock - nothing in my argument hangs on this support. The main point I wish to make from this consideration of ancient views, is that there is a long tradition for conceiving time as ontically prior to the existence of matter and its changes. This formulation goes all the way back to Pherecydes, at least on the interpretation of his words given above, who asserted that time must always have existed since matter existed even before Zeus gave it form, and it existed in time. At any rate, after citing Pherecydes' view Aristotle remarks: "It is just this that enables Democritus to show that all things cannot have had a becoming; for time, he says, is uncreated."[15] On Democritus' view, then, not everything that exists is material; matter exists in the void, and both exist in uncreated time. Thus for the first systematic materialist, time is ontically prior to change, even if it is not a material being. This standpoint is in marked contrast to the Epicureans, for whom time is an accident of motion, which in turn is an accident of body, yielding a status for time as "an accident of accidents". This

---

[10] Aristotle, *Physics*, 219b 2-4.
[11] *ibid*, 223a 24-25.
[12] *ibid*, 219a 9.
[13] *ibid*, 223a 27-28.
[14] *ibid*, 218b 12-13. "The reason," Aristotle claimed, "why people think of time as the change of the heavenly sphere is that all other changes are measured by this change, and time is too." (*Physics*, 223b 21-23). He also argued that "all changes may be faster or slower, but not so time; for fast and slow are defined by time, 'faster' being more change in less time, and 'slower' less in more."(*Physics*, 218b 14-16). This same argument Leibniz was to try some two thousand years later against Newton's "equable flow of time" in his controversy with Clarke - but without much success.
[15] Aristotle, *Physics*, 251b 15-17.

standpoint is the second major type of materialist theory of time, about which I shall have more to say later.

Democritus' conception, rather than Epicurus', was taken up by the Stoics, who form an important bridge from ancient conceptions to the seventeenth century. For them, time is an infinite container analogous to the void. Apollodorus, for example, defined time as an *infinite interval*,[16] in which respect he followed Chrysippus:

> It is in Time that everything moves and exists. ... Just as the void in its totality is infinite in every respect, so time in its totality is infinite on either side; for both past and future are infinite.[17]

The significance of the Stoics in this connection is that this notion of place and time as infinite intervals was taken up by Pierre Gassendi in the seventeenth century, and, as I have shown elsewhere,[18] Gassendi's views were a major source for Newton. Decidedly, Gassendi was a thorough-going materialist - or, at least, as thorough-going a materialist as a man of the cloth in seventeenth-century France could be.[19] He maintained the materiality of the soul, at least the vegetative and animal ones, exempting only the rational soul. Celebrated as the chief reviver of Epicureanism, he promoted a naturalistic ethics that in many ways anticipated the utilitarian theories of Mandeville and Helvétius. However, on space and time, he chose to follow the Stoics rather than Epicurus. After commending Chrysippus for his comparison of time with the void "especially with regard to their infinity", he attempted a "more explicit explanation of the matter" as follows:

> As Place as a whole is unlimited, so Time as a whole has neither beginning nor end, and as any moment of Time is the same in all places, so any part of Place is in all times. Likewise as Place persists immobile whether anything exists in it or not, so Time flows with an equal tenor whether anything endures in it or not,

---

[16] Apollodorus: Time is "the interval of the world's motion, infinite in just the same way that the whole of number is said to be infinite." in Long, A. A. and Sedley, D. N., *The Hellenistic Philosophers*, vol. 1 (Cambridge: Cambridge University Press, 1987), pp. 304-305.

[17] Stobaeus 1.106,5-23 (SVF 2.509); partly quoted from Sambursky in Čapek, Milič (ed.), *Concepts of Space and Time* (Dordrecht/Boston: D. Reidel, 1976), pp. 160-161); and partly from Long and Sedley, *op. cit.*, p. 304.

[18] Arthur, Richard T. W., "Newton's Fluxions and Equably Flowing Time", *Studies in History and Philosophy of Science*, 28, no. 2, (June 1995): 323-351.

[19] Despite his friendship with Hobbes and his outspoken defence of the ancient atomists, and his having among his protégés such libertines as the swash-buckling Cyrano de Bergerac, Gassendi was sufficiently careful not to have to flee his country, as did Descartes, Bayle, Hobbes and Locke. For a reading of Gassendi as a thorough-going materialist, see Bloch, Olivier, *La philosophie de Gassendi* (Dordrecht: Springer, 1971).

> whether anything is at rest or in motion, whether it moves faster or slower. [20]

Pursuing this analogy with space or void, Gassendi quotes Seneca on Plato's distinction of all entities into six classes, the sixth of which (in Walter Charleton's translation) "contains only those things which have *as it were* a being, as INANITY and TIME: which we thus expound, *Space and Time are things more General then to be comprehended under the Categories of Substance and Accident.*" Or, as Charleton says more colourfully in his heading for this article, "Time, nor substance, nor Accident: but an *Ens* more General, and the Twin-brother of *Space*."[21] Newton amplifies this same Platonic theme:

> Concerning extension, perhaps it may now be expected that I should define it either as a substance, an accident, or else nothing at all. But by no means, for it has a certain mode of existence proper to itself which is fitting for neither substances nor accidents. It is not a substance; on the one hand because it does not subsist absolutely *per se*, but, as it were, as an emanative effect of God, and a kind of affection of every being; and on the other hand because it is not among the proper affections that denote substance, namely actions, such as thoughts in the mind and motions in a body.[22]

On this interpretation, space and time, although neither substances nor accidents, exist independently of the matter in them. We may note in passing that it therefore seems to be a mistake to classify Newton's view as "substantivalist", as is usually done, a label perhaps more apposite for the early physicists referred to by Aristotle. In fact, it might have been more accurate to call this an "existentialist" account, if Sartre had not appropriated the term. For, as Newton continues:

> Space is an affection of a being insofar as it is a being. No being exists or can exist which is not related to space in some way. God is everywhere, created minds are somewhere, and body is in the space that it occupies; and whatever is neither everywhere nor anywhere does not exist. And hence it follows that space is an

[20] *Animadversiones*, quoted from Čapek, *op cit*, p.199.

[21] Charleton, Walter, *Physiologia Epicuro-Gassendo-Charltonia: or a Fabrick of Science Natural upon the Hypothesis of Atoms*, (c.1654), Reprinted in *Sources of Science*, 31 (N.Y. London: Johnson Reprint Co., 1966), p.75.

[22] This and the next translation from Newton are my own; I have referenced them to those in Janiak, Andrew (ed.), *Newton's Philosophical Writings*, (Cambridge: Cambridge University Press, 2004), p. 21.

> emanative effect of the first existing being, since if any being whatsoever is posited, space is posited. And the same may be asserted of duration: for certainly both are affections or attributes of a being according to which the quantity of each individual's existence is denominated with respect to amplitude of presence and to persistence in its being.[23]

As we have seen, this view of time as uncreated and coeval with God goes back not just to the Stoics, with their view of the void and time as uncreated, infinite intervals, but to Democritus and even to the ancient Ionians.

In modern times, however, most materialist accounts of time have proceeded in a diametrically opposite direction to that taken by the Stoics and Gassendi. Epicurus, as we have seen, held the more Aristotelian view that time is an accident of motion, itself an accident of material beings, and I shall have occasion to return to this below. Hobbes took an even more radical view, holding that time is a "phantasm of motion". On the one hand, this view is a result of his nominalism: abstract ideas like space and time are formed by abstraction from, or reflection on, particulars. On the other hand, though, the term "phantasm" is of course a term of art for Hobbes. In his psychologized epistemology, the phantasm of space -"out-thereness", Kant's "form of outer sense"- is formed in the brain as a reaction to the endeavours (another term of art) that are imposing themselves on the sense organs from without. Space is a physical reaction, a counter-endeavour by the organs (particularly the eyes), subjectively apprehended. (This is part of Hobbes' ingenious double-aspect solution to the mind-body problem, that so enthused the young Leibniz. The same endeavour that is (taken objectively) a vanishingly small motion of the subject towards or away from an object, is (taken subjectively) an appetite or aversion for that object). Similarly, the passing of time is a phantasm in the brain caused by the phenomena of motions themselves, another counter-endeavour, subjectively apprehended.

Hobbes once remarked (approvingly) of Spinoza that "he has out-thrown me a bar's length" - although we have no record of the bar in question - "for I durst not write so boldly".[24] This was in reference to Spinoza's theology and politics,[25] although it is also perhaps applicable to their views on time, where Spinoza goes beyond Hobbes in stressing the derivative nature of time. In Spinoza's view, it is duration that is fundamental. One could argue that he inherits this from Descartes, for whom time is a *quantity*,

---

[23] *ibid*, p. 25.

[24] A remark of Hobbes's reported by John Aubrey in Clark, Andrew (ed.); *"Brief Lives," chiefly of Contemporaries, set down by John Aubrey, between the years 1669 & 1696*, 2 vols., (Oxford: Oxford University Press), vol.1: p. 357.

[25] For an analysis, see Edwin Curley's "Religion and Morality in Hobbes" in Coleman, Jules and Christopher Morris (eds.); *Rational Commitment and Social Justice*. (Cambridge: Cambridge University Press, 1998).

whereas duration is something only conceptually distinguished from existence. Newton, as we have seen, equates duration with quantity of existence in respect of something's persistence in its being.[26] In so doing, he collapses the Cartesian distinction, regarding duration as another name for time. Spinoza, on the other hand, upholds the distinction, but demotes time to a lesser status, that is nothing but a *mode of thinking* or rather of "imagining".[27] It arises, he explains, when we try to abstract quantity from substance and "separate duration from the way it flows from things eternal".[28] Substance, that is, has an eternal existence, whereas the existence of its modes, including bodies, "is explained by means of duration". "Time", on the other hand, "is for limiting duration, measure for limiting quantity, so that we may as far as possible to imagine them easily."[29] Thus, time is the result of the attempt to abstract quantity from duration, the abstraction being a mental operation that he regards as intrinsically flawed. So we see in Spinoza the kind of nominalistic trend we found in Epicurus and Hobbes, but with an anticipation of Hegel: time is an abstraction from duration, but abstraction is a kind of falsification.[30]

Materialism

So far, dealing with the history of philosophy, I have been able to take for granted a certain rough and ready notion of what materialism is. Now I want to say something by way of characterizing it. In the twenty-first century, materialism is usually equated with physicalism: everything that exists is made of matter and the various forms it takes; all existing things, processes and phenomena can be explained as manifestations or results of matter. This definition is reasonable enough as a first pass, allowing some license in the definition of matter, and in what it means to be a result of matter. In this sense, materialism is not so much a theory with precise and refutable content, but is better regarded as a research program, in Lakatos' sense. It is a set of background assumptions (the "hard core") that lies behind most scientific research and guides it as to what hypotheses are reasonable to make. It is not considered refuted, for example, by the discovery that electromagnetic and

---

[26] Parenthetically, it has been argued in a forthcoming article by Steffen Ducheysne that another significant influence on Newton here is Spinoza's contemporary van Helmont.

[27] Spinoza, "Letter of the Infinite", quoted from Leibniz's transcription in: Leibniz, Gottfried W., *The Labyrinth of the Continuum Writings on the Cotninuum Problem, 1672-1686* 2nd edition. selected, translated and edited by Richard T. W. Arthur (New Haven and London: Yale University Press, 2001), p.109.

[28] *ibid.*

[29] *ibid.*

[30] This interpretation is no doubt a somewhat tendentious reading of Spinoza, as Peter Loptson reminded me. One could also see him as resounding Platonic and Neoplatonic themes (as do also Newton and Leibniz, in different ways). For the roots of the distinction between *aeonic* and *chronological* time in Plato, see Kavanaugh, Leslie Jaye, *The Architectonic of Philosophy: Plato, Aristotle, Leibniz* (Amsterdam: Amsterdam University Press, 2007), esp. p. 41.

other fields can exist in empty space without modifying some material ether, or by the interconvertibility of matter and energy discovered by Einstein, or by the unexplained existence of dark energy; rather, the definition of matter is revised to accommodate such counterexamples. Provided this can be done in a progressive way rather than by *post hoc* counter-moves, the defense of materialism as a research program is viable.

But without further elaboration, the above definition makes it still a mechanistic or Cartesian materialism, of the kind entertained by Descartes' disciple LeRoy (Regius) and also in the following century by La Mettrie. On that philosophy, matter is merely passive, defined in opposition to active spirit; such a materialism is therefore vulnerable to the objection that it is extremely hard to see how active things - minds and spirits - can "result from" matter that is passive, merely *acted upon* by other bodies and not *acting* in any authentic sense. It was the inadequacy of this conception that opened the door to the idealism of Fichte and Hegel. As a matter of historical fact, such objections (already made by Bishop Bramhall to Hobbes and by Bishop Wilberforce to Locke, as well as by Bishop Berkeley) were countered by conceptions of matter as *inherently active*: Toland in Ireland, Priestley and Hartley in England, and Buffon and Diderot in France. I believe that this was an essential movement of materialist thought in clearing the ground for Darwin, although I cannot make that case here. In any case, it seems to me that the definition of materialism as physicalism is deficient in that it requires supplementing by some reference to the inherent disposition of matter to act - in a sense that goes beyond its merely being set in motion, for that would seem to require an external agent. Otherwise the emergence of higher order forms by evolution is left without adequate grounding. So, as a supplement to the definition of Materialism: all existing things, processes and phenomena can be explained as manifestations or results of *the actions and interactions of material things, with these actions and interactions having their origin in matter's inherent disposition to act.*

Furthermore, I would also dispute the adequacy of the physicalist definition for a different reason; it does not capture another important strand of materialist thought, that which was initiated by Ludwig Feuerbach, and extended by Engels and Marx. Marx Wartofsky, in his excellent study of Feuerbach, [31] comments on this very eloquently. After noting how Feuerbach's anthropological materialism is typically represented as reducible to his epigram "Man is what he eats", and Marx's historical materialism likewise dismissed as an economic reductionism, he writes:

> The question "What is materialism?" goes beyond ... such simpleminded formulations. ... To talk of "material human

[31] See Marx Wartofsky, in his excellent study of Feuerbach. Wartofsky, Marx W., *Feuerbach* (Cambridge: Cambridge University Press, 1977).

> existence" is to talk of neither atoms and molecules, nor blood, cells and organs (including the brain); rather, it is to talk of social and historical human existence. But in what sense is this "material"? Is the metaphysical extension from physical "matter" inevitably reductive? Does emergentism provide anything but the pious accommodation of a programmatic materialism, and cover up rather than reveal the problems raised by the claim for continuity and unity between the natural, the social and the mental?[32]

I am not going to attempt to resolve such questions here, but I do think one can identify a commonality between these various forms of materialism that goes beyond the minimal physicalist core. Let me begin with Darwin, with whom one finds a quite self-conscious commitment to materialism in his early notebooks: "It is an argument for materialism, that cold water brings on suddenly in [the] head a frame of mind analogous to those feelings which may be considered as truly spiritual." [33] Darwin goes on to argue, "[a]ffections [are the] effect of organization, which can hardly be doubted when seeing Nina with her puppy…[T]hinking over these things, one doubts [the] existence of free will. [E]very action [is] determined by hereditary constitution." [34] Nevertheless, "to avoid stating how far I believe in Materialism, say only that emotions, instincts[,] degrees of talent, which are hereditary, are so because brain of child resembles parent stock."[35] Darwin concludes: "Now is it not a little remarkable that the fixed laws of nature should be universally thought to be the will of a superior being, whose natures can only be rudely traced out. When one sees this, one suspects that our will may arise from as fixed laws of organization."[36]

The tendency of all these speculations is to give naturalistic explanations of things previously attributed to non-material agencies: affection and other emotions and free will as attributes of the mind, divine will as an attribute of God. In this, Darwin is basically following the model inaugurated by Epicurus and Hobbes, in which materialism has an essential liberating and political component in freeing people from a belief in entities by which they have been held in thrall. This aspect of materialist thought is well described by Douglas Jesseph in the course of making the case for Hobbes' materialism as committing him to atheism: "Hobbes held that one of the main goals of the true philosophy is to overcome our ignorance of the

---

[32] *ibid*, p.xii.
[33] Darwin, Charles, *Metaphysics, Materialism and the Evolution of Mind: Early Writings of Charles Darwin*, transcribed and annotated by Paul H. Barrett (Chicago: University of Chicago Press, 1974), p.9.
[34] *ibid*, p. 10.
[35] *ibid*, p. 16.
[36] *ibid*, p. 14.

world's workings—a process that involves discarding many of the entities invoked by earlier generations to explain the phenomena of nature."[37]

Thus, a distinctive characteristic of materialist thought is a certain kind of *nominalism*, in which entities mistakenly thought to exist independently of us and our "temporal" activity are seen instead to be artefacts resulting from interactions among material things, organisms, or people. In so identifying nominalism as a central component of materialism, I follow Engels and Marx; although where they see traditional medieval nominalism as "the first form of materialism", I rather see a certain form of it as the engine driving materialist thinking.[38] Commonly, nominalism in its general acceptation, denotes a position arising in a medieval dispute over abstract terms or universals, such as "humanity". A nominalist would deny that it is necessary to posit the existence of entities corresponding to such terms, "universals", over and above the individuals or particulars that exemplify the property in question. I am extending the meaning of this "entity-skepticism" to things other than abstract terms, to any universal non-material beings that are thought to be the sources of actions affecting us: Eros and other divine powers, the faculty of free will, a designer God, and so forth. Yet it is not to be thought that the effects so described are unreal, just that they are wrongly attributed to general, non-material entities, rather than particular material individuals. I would gladly cede the term "nominalism" to its normal meaning, if I could find a better word to express this kind of philosophical tendency. But for want of a more apposite term, from now on when I say "nominalist", I mean the tendency I have just described, rather than the position usually associated with Willard Quine and Nelson Goodman. One can see a similar nominalist trend of thought in Feuerbach (1804–1872), who was an exact contemporary of Darwin's (1809–1882). As Wartofsky reports:

> Feuerbach sees the concept of an absolute Will as the creation of human imagination, and as an incursion upon the principle of reason. In this respect, says Feuerbach, Spinoza is the true founder of modern philosophy, in that he explicitly rejected such a notion and systematically explained its origin in human feeling and in the

---

[37] Jesseph, Douglas, "Hobbes' Atheism", *Midwest Studies in Philosophy*, *XXVI*, pp. 140-166. (2002): 162.

[38] Thus in *The Holy Family*, Marx quotes Engels from the introduction of his *Socialism: Utopian and Scientific*. See Engels, Frederick, *Ludwig Feuerbach and the Outcome of Classical German Philosophy*, ed. C. P. Dutt (New York: International Publishers, 1941), pp. 10-12. Also, "Nominalism, the *first form* of materialism, is chiefly found among the *English* schoolmen" in Marx, Karl and Friedrich Engels, *The Holy Family; or Critique of Critical Criticism*, transl. Richard Dixon and Clemens Dutt (USSR: Progress Publishers, 1975), p. 158.

> imagination as an anthropomorphic projection of human will and desire.[39]

It is in this systematic explanation of entities such as the Will as anthropomorphic projections that Feuerbach's materialism goes beyond that envisioned by Darwin and Hobbes, where materialism simply *discards* these entities. This notion of anthropomorphic projection is further developed by Feuerbach in his analysis of God: "Man --this is the mystery of religion-- objectifies his being and then again makes himself an object to the objectivized image of himself thus converted into a subject."[40] Feuerbach goes on to state:

> Religion is man's self-estrangement: man posits God over against himself as a separate being. ... It must therefore be shown that this opposition, this dichotomy of God and man, which is the starting point of religion, is a dichotomy within man, of man with his own essence.[41]

As is well known, this Feuerbachian analysis of religion was hugely influential on Engels and Marx, who applied a similar critique in economics. As Engels wrote: "Economics deals not with things but with relations between persons, and, in the last resort, between classes; these relations are, however, always *attached to things* and *appear as things*."[42] Marx, for instance, applies a parallel argument to Feuerbach's in his doctrine of the fetishism of commodities, which are experienced as entities with their own values, but according to Marx instead arise out of material, social relations that are then objectified and treated as governing our behaviors.

I will not go any further into the history of materialism. But perhaps I have done enough to suggest that missing from the equation with "physicalism" is everything that has made materialism a dangerous doctrine to subscribe to: its theological and moral subversiveness, and consequent political radicalism. This missing ingredient I identify, following Engels and Marx, is a species of nominalism: entities mistakenly thought to exist independently of us and our "temporal" activity - whether souls (la Mettrie), religion-based moral law (d'Holbach), fixed biological species (Diderot, Darwin), gods (Feuerbach), or commodities (Marx) - are revealed to be artefacts resulting from interactions among material things, organisms, or people; and in certain cases these entities can be revealed to be

---

[39] Wartofsky, *op cit,* p.115.
[40] *ibid.*
[41] Feuerbach, (VI, 41; S: I, 81). quoted from Wartofsky, *op cit,* p. 293.
[42] Engels, Friedrich, "Karl Marx's *Contribution to the Critique of Political Economy*" in Engels, Frederick, *Ludwig Feuerbach and the Outcome of Classical German Philosophy*, ed. C. P. Dutt (New York: International Publishers, 1941), p. 80.

objectifications of human, social relations that are then posited as existing apart from us and governing our lives.

> Nominalist Supplement to the Definition of Materialism: *Entities mistakenly thought to exist independently of us and our "temporal" activity are revealed to be artefacts resulting from interactions among material things, organisms, or people; and in certain cases these entities can be revealed to be objectifications of human, social relations that are then posited as existing apart from us and governing our lives.*

Armed with this take on materialism, let me return to the topic of time, with an eye now to contemporary theory of time.

The "Elimination of Time"

Today one of the strongest new initiatives in understanding time has come from an investigation into the deep metaphysical obstacles to erecting a theory of quantum gravity. The problem, in a nutshell, is a deep-rooted tension between the way time is treated in each of the revolutionary theories of the early twentieth century. To formulate the problem briefly: in relativity theory spacetime is fundamental, and time is as it were, a mere aspect of this structure. Both space and time are warped by high concentrations of mass-energy. But according to quantum theory, for sufficiently small "times", the mass-energy spread is exceedingly large, so that - given the warping of spacetime by mass - the concepts of spacetime location and duration will lose all determinateness. A second notable problem is that whereas time and spacetime are dynamic in General Relativity, and tied to the distribution of mass-energy, in quantum theory time appears as a mere parameter. The upshot of this situation is that most theorists agree that a search for a viable theory combining the two - a theory of quantum gravity - will likely involve a profound revision of classical ideas about space and time, going beyond even the revolutionary changes in our understanding of time wrought by Einstein's relativity theories.

In recent years, new life has been injected into the quest for a theory of quantum gravity from an unexpected quarter: an independent researcher, Julian Barbour, working outside the academy (and supporting himself by translating Russian), wrote a massive tome tracing the debate between absolutist and relativist conceptions of motion, supporting the relativist line taken by Leibniz and Mach.[43] Against Newton's conception of motion relative to absolute space, Leibniz had objected that if one supposed the

[43] Volume 1, alone, of Barbour's *Absolute or Relative Motion* numbers 746 pages, and he does not even get to Leibniz or Mach! Barbour, Julian, *Absolute or Relative Motion? Vol. 1: The Discovery of Dynamics* (Cambridge: Cambridge University Press, 1989).

world and everything in it, while maintaining the same mutual spatial relations, were moved in a straight line relative to such a space, there would be no way even in principle to determine the supposed change, making the supposition of an absolute motion "chimerical". "For two states indiscernible from each other are the same state, and consequently 'tis a change without any change."[44] But this concerns only unaccelerated motion in a straight line, which Newton had acknowledged would be undetectable. Leibniz does not address Newton's contention that absolute motion could nevertheless be discerned in the centrifugal effects seen in a rotating bucket of water. Ernst Mach was left to counter Newton's contention that this is evidence for absolute motion:

> Newton's experiment with the rotating vessel of water simply informs us that the relative rotation of the water with respect to the sides of the vessel produces *no* noticeable centrifugal forces, but that such forces *are* produced by its relative rotation with respect to the mass of the earth and the other celestial bodies.[45]

Now if Barbour were just another proponent of the relativist cause on philosophical grounds, this would be no cause for celebration. But following on his historical research, he was able, in collaboration with Bruno Bertotti, to supply a mathematical demonstration of Mach's claim (which Mach had made on purely epistemological grounds) that inertial motion could be represented as a purely relative motion without presupposing Newton's absolute space and time: "Just as Mach had suspected, the phenomenon that Newton called inertial motion in absolute space could be shown to arise from motion relative to all the masses in the universe."[46] What is more, Barbour and Bertotti showed that a similar approach, based on only relative motions, could be taken in General Relativity, but that when they did this, surprisingly, they recreated Einstein's theory by another route. The effect of this was to give an arsenal of ammunition to those who were not convinced that the lesson of General Relativity was that spacetime had to be treated as an entity in its own right.

The connection of all this with time is given by another Machian dictum: "It is utterly beyond our power to measure the changes of things by time. Quite the contrary, time is an abstraction, at which we arrive by means

---

[44] 4th paper, §13. in Leibniz, Gottfried W., *Philosophical Papers and Letters*, 2nd ed., selected, translated and edited by Leroy Loemker (Dordrecht: D. Reidel, 1976), p.688.

[45] Mach, Ernst, *The Science of Mechanics: a Critical and Historical Account of its Development* (c.1893), (La Salle, Illinois: Open Court, 1974), quoted in Huggett, Nick, *Space from Zeno to Einstein* (Cambridge: MIT Press, 1999), pp. 176-177.

[46] Barbour, Julian, *The End of Time: The Next Revolution in Physics* (Oxford: Oxford University Press, 1999), p.114.

of the changes of things."[47] In order to give this idea flesh, Barbour and Bertotti define time in terms of what they call the "Machian distinguished simplifier", a kind of average of all the changes occurring in the universe. Barbour concludes: "Time is change, nothing more, nothing less."[48] This relativist perspective has paved the way for similar relativist approaches to General Relativity by leading quantum gravity researchers such as Lee Smolin and Carlo Rovelli. They interpret time not as an aspect of a pre-existing spacetime structure, but as a concomitant of a purely relational dynamics from which time and spacetime emerge as results.

Barbour and Rovelli both express this conception as resulting in the "elimination of time", while Barbour and Smolin identify Leibniz as their chief inspiration. There are two main points I wish to make about this situation. First, to me, Barbour's Machian initiative seems to fit squarely in the tradition of materialist thinking on time, even though some of the specifics of his views are Platonic. Second, I agree about its Leibnizian provenance, with the result that Leibniz, perhaps surprisingly, appears as the crucial link here between Spinoza and Hobbes on the one hand, and the materialist conception that I see in Barbour's Machian relativity on the other.

Initially, on the status of time, Rovelli expresses himself unequivocally as follows: "The disappearance of physical time is the second characteristic feature of the relativistic revolution."[49] Rovelli goes on to the position in which "I consider the viability of a foundation of our understanding of the world in which space and time play no role."[50] His editor, Dennis Dieks, perhaps to protect him from idealist misconstrual, explains that "[what Rovelli] means is that space and time no longer enter as independent entities, on top of what is already determined by all the coincidence relations between the dynamical fields."[51] If Dieks is right (and I think he is), then Rovelli's position seems much more like the position we have encountered in Epicurus, Hobbes and Spinoza. Time is not an existing entity; rather it is an aspect of motion and change. This is also in keeping with Mach's dictum quoted earlier, "time is an abstraction, at which we arrive by means of the changes of things". And in fact Barbour himself writes: "Time does not exist. All that exists are things that change. What we call time is - in classical physics at least - a complex of rules that govern the change."[52] But this is not just a view that is in agreement with some materialist thinkers - others have taken the view that Rovelli and Barbour are opposing, as we have seen - it is in keeping with the kind of nominalist tendency that I have

---

[47] *ibid,* p. 67.
[48] *ibid,* p. 120.
[49] Rovelli, Carlo, "The Disappearance of Space and Time", pp. 25-36, in Dieks, Dennis, (ed.), *The Ontology of Spacetime* (Amsterdam: Elsevier, 2006),p. 32.
[50] *ibid,* p. 25.
[51] Dieks, *op cit,* p. xii.
[52] Barbour, *The End of Time,* p.137.

identified as the engine of materialist thought: entities that were thought to be self-existent turn out, on a deeper analysis, to be aspects of relations among things, relations which were previously mistaken for relations between that entity, taken as self-existing, and its manifestation in those things.

This characterization is not, however, how Barbour situates his own theory. He sees the materialists as typically espousing absolutism about space, time and motion, and relativism as being the preserve of idealists.[53] Moreover, for him, the elimination of time is also the elimination of becoming and change, in agreement with the Parmenidean idealist tradition: "I also think that Plato was right when he said that Being...is real, and Becoming is an illusion....Platonia is the arena that I think must replace space and time....The points of Platonia -- the Nows -- are worlds unto themselves. No thread of time joins them up."[54]

Naturally, Barbour also takes this to be consonant with his debt to Leibniz. For just as Leibniz denied that absolute space is an entity, so he also denied the same of time: "space in itself is an ideal thing like time" (to Clarke, 5th Paper, §33; 1976, 701). But about time Leibniz also says: "Nothing of time ever exists except instants, and an instant is not even a part of time. Anyone who considers these observations will easily comprehend that time can only be an ideal thing."[55]

Leibniz is interesting in that Barbour's brief for the elimination of time is that the world (his "Platonia"), is best conceived as being simply the collection of all possible "nows" or instants. And these instants are conceived, as by Leibniz, as orders of possible co-existents, configurations which he also calls "time capsules":

> [Lee Smolin and I] met several times in the next few years, and collaborated on an attempt to formulate Leibniz's philosophical system, his "monadology", in mathematical form....Certain aspects of our work together were decisive in my own elaboration of the notion of time capsules and my conviction that the ultimate and only truly real things are the instants of time.[56]

Barbour is explicit in proposing a nominalistic basis for this view: space and time are not basic entities, but derived from more basic things:

---

[53] Barbour notes that "broadly speaking, the materialists (or realists) were identified with the absolute (Newtonian) approach to motion, the idealists with the relationals (Machian) conception, although there were notable exceptions." Barbour, *Absolute or Relative Motion,* p. 15.

[54] Barbour *The End of Time*, p. 45.

[55] Letter to Clarke, 5th Paper, §49) in Leibniz, *Philosophical Papers and Letters*, *op cit,* p. 705. See also, Arthur, Richard T. W., "Leibniz's Theory of Time," in Okruhlik, Kathleen and Brown, James R. (eds.). *The Natural Philosophy of Leibniz* (Dordrecht: D. Reidel, 1985), pp. 263-313.

[56] Barbour *The End of Time*, p. 240.

> We have been exploring Leibniz's idea that only things exist and that the supposed framework of space and time is a derived concept, a construction from things. If it is to succeed, the only possible candidates for the fundamental "things" from which the framework is to be constructed are configurations of the universe: Nows or "instants of time". They can exist in their own right: we do not have to presuppose a framework in which they are embedded. In this view, the true arena of the world is timeless and frameless - it is the collection of all possible nows.[57]

Obviously, it is beyond the scope of his paper to give an adequate exposition of the whole of Barbour's argument for this position, or even an analysis of the profound problems besetting any philosophical position that denies the reality of change. But I do want to address the debt to Leibniz, for here I think that Barbour has matters the wrong way up. For instants are precisely not *things* for Leibniz, but rather abstractions from relations among things. Yet although Leibniz does not at all follow Spinoza in seeing abstraction as falsification, but sees it rather as a necessary condition for doing mathematics, still, an abstract entity does not need to be posited as an existing thing. At the basis of Leibniz's nominalism is a deep commitment to individual substances: these are his "true things". On the other hand, his monads are not changeless Platonic forms, but instantiated individuals whose changes are manifested in the changing phenomena resulting from them. Leibniz is explicit on this point in his correspondence with De Volder in 1703:

> You doubt, distinguished sir, whether a single simple thing would be subject to changes. But since only simple things are true things, the rest being only beings by aggregation and thus phenomena, and existing, as Democritus put it, by convention and not by nature [*nómo* not *phúsei*], it is obvious that unless there is a change in the simple things, there will be no change in things at all.[58]

Thus, all things change, and composite things (phenomena) change because of the changes in simple things from which they result. Instants, meanwhile, are abstractions from sets of changes of state that are happening simultaneously, while the relation of simultaneity is further analyzed in terms of the compatibility of the states of the substances. The ideality of instants, like that of parts of time and place, applies to them as abstractions from these changes:

---

[57] *ibid,* p. 177.
[58] Leibniz, *Philosophical Papers and Letters*, *op cit,* p. 531.

> The parts of time and place, considered in themselves, are ideal things; and therefore they perfectly resemble one another, like two abstract units. But it is not the same with two *concrete ones*, or with two *real times*, or two places *filled up*, that is to say, truly *actual*.[59]

Places and times, that is, can be *concreta*, if they are instantiated by actual things and their changes. With regard to space, Leibniz gives a thorough account of this form of nominalism in his Fifth Paper to Clarke (§47), explaining the difference between *concrete space*, which he later (§104) describes as the "order according to which situations are disposed", and *abstract space*, which is "that order of situations when they are conceived as being possible," and which "is therefore something ideal."[60]

On the other side of this coin, Leibniz's substances always have a situation through the phenomenal body of which they are the entelechy or soul:

> Every change, spiritual as well as material, has its own seat [*sedes*], so to speak, in the order of time, as well as its own location in the order of co-existents, or in space. For although monads are not extended, they nevertheless have a certain kind of situation in extension, that is, they have a certain ordered relation of co-existence with others, namely, through the machine they control. I do not think that any finite substances exist apart from a body, so that they would lack a position or an order in relation to the other things co-existing with them in the universe.[61]

That is to say, because substances are individual things that act, and the actions are only instantiated in the orders constituting (concrete) space and time, Leibniz's monads are necessarily embodied. This position is a very far cry from the world as conceived by Barbour, where not only space and time but even individual, acting things are none of them part of *what is*, but are all, in some difficult sense, resultants of the nexus of possibles and the way we experience it. Thus insofar as Barbour, along with Smolin and Rovelli, wants to treat time not as an entity, but as an aspect of reality deriving from motion and change, he is proceeding in keeping with Leibniz - and with the philosophies of Epicurus, Hobbes and Spinoza. But the moment he goes beyond this with his philosophy of the instants of time as "the ultimate and only truly real things", he has completely departed from the kind of nominalism that Leibniz espouses, and gone to Platonia.

---

[59] Letter to Clarke, V, §27 in *ibid,* p.700.
[60] *ibid.*
[61] Letter to de Volder, June 20, 1703,) Leibniz, *op cit,* p. 531.

Leibniz and Materialism

We have seen that Leibniz is committed to a certain realism about material bodies and their changes, so long as they are actualized by the monads responsible for their acting. Because of the dominance of idealistic interpretations of his thought, this feature of Leibniz's philosophy has not been widely recognized. Leibniz's commitment to a kind of body realism was the subject of a joint paper by Peter Loptson and myself, in which Peter suggested a different interpretation of this body realism than the one I am suggesting here. But in closing, I want to turn to a consideration of Leibniz's relationship with materialism. For although Leibniz is opposed to the kind of Cartesian materialism I outlined earlier, it should not be thought that he advocated in its stead a pure immaterialism. As he writes to Sophie Charlotte:

> When it is said that there are immaterial substances, one means by this that there are substances which include other concepts, namely, perception and the principle of action or change, which cannot be explained either by extension or impenetrability....Hence if anyone says that force and perception are essential to matter, he is taking matter for the complete corporeal substance which includes form and matter, or the soul along with the organs. This is the same as if he had said that there are souls everywhere. This could be true, yet not at all contrary to the doctrine of immaterial substances.[62]

Thus, we might say with Feuerbach, Leibniz's idealism is "compromised" by a materialism that has not been properly eradicated. Although what exist are immaterial substances and their forces and perceptions, these forces can only manifest themselves in bodies, and perceptions are likewise of bodies and depend on bodily organs. Material substances are necessary concomitants of (an infinite plurality of) immaterial ones.

My mention of Feuerbach in this connection will help me to close the circle of my argument. For Feuerbach gave a very perceptive critique of Leibniz in 1837, at a time when he was still an enthusiastic Hegelian idealist. He lauds Leibniz for his recognition that matter cannot be purely passive, but must contain principles of activity. As Wartofsky says, Feuerbach is always opposed to Cartesian materialism.[63] But in the course of penning the critique, he begins to convert himself to materialism. The key thing he learned from Leibniz in this transformation, by his own account, was the existential primacy of the individual - situated in its own particular setting - over the abstract. For, according to Wartofsky, the two shortcomings Feuerbach identifies in Leibniz's idealism are the "mere" individuality of substances and a compromising materialism in that the monads are only able to perceive

[62] Leibniz, *ibid,* pp. 551-2.
[63] Wartofsky, *op cit*, p.76 *ff.*, pp. 97-99.

each other imperfectly through the "veil of matter". As Wartofsky notes, ironically, "Feuerbach chooses precisely these two elements that he criticizes in Leibniz as the cornerstones of his own later empirical materialism."[64]

So this is my concluding suggestion: precisely these same features of Leibniz's thought that set Feuerbach on the right road to materialism, explain how Leibniz's theory of time could be a bridge from Hobbesian and Spinozist materialism to the materialist theories of time being entertained in the search for a theory of quantum gravity.

---

[64] Wartofsky, *ibid,* p. 103. I think this is no accident, but stems from his recognition that Leibniz's conception of activity is effete, alienated.

Corollaries on Space and Time: A Survey of Arabic Sources in Science and Philosophy [1]

Nader El-Bizri

This chapter examines selected theories of space and time in classical traditions in science and philosophy within the history of ideas in Mediaeval Islamic civilization. Reflections on the essence and existence of space and time preoccupied scientific and philosophical thinking since its earliest foundational epochs. The adaptive assimilation, critical interrogation, and innovative expansion of classical Greek traditions in science and philosophy informed the scholarly debates in mediaeval Arabic sources on space and time. Some wondered whether time was altogether nonexistent, while others doubted the reality of its divisibility into parts by arguing that the past ceased to be, that the future does not yet exist, and that the present as a moment/now, which is without magnitude, would not constitute a real part of time. The physical definition of place was also challenged by way of positing place as geometric space. The question concerning the essence and existence of space and time carried significant metaphysical and cosmological entailments that animated the debates between the philosophers (exponents of f*alsafa*) [2] and the dialectical theologians (proponents of *kalam*). Theological beliefs in the temporal origination of the universe by way of creation and opposing philosophical doctrines of the eternity of the world were also entangled with ontological reflections on the reality of nothingness and the existence of the void, versus the positing of space as a virtual vacuum or a postulated emptiness. Such corollaries on space and time were ultimately central to mediations on divinity when thinking at the "limits of human understanding".

Prologue

In this survey, I shall present some of the principal theories from Arabic mediaeval sources in the exact sciences and philosophy, regarding the essence and existence of space and time, while I shall also give a succinct account of the main classical Greek traditions that received their adaptive commentaries, solicited their reforming critiques, and inspired their inventive initiation of novel directions in thinking.

Space: Classical Conceptions of Space

The question concerning the reality of space, its specific kind of being and quiddity, has been debated by scholars since the foundational unfolding of philosophical thought. As Aristotle noted in book *Delta* (IV) of his *Physics*, numerous classical thinkers endeavored to affirm the *existence* of place

---

[1] It is worth noting from the onset, that this title refers to "Arabic" as the *lingua franca* of classical traditions in science and philosophy of mediaeval Islamic civilization; it is not meant to indicate that scholarship in this intellectual *milieu* was primarily and solely associated with the Arabs, given that many thinkers were Persian and Turkish. Moreover, while the majority of the scholars of mediaeval Islamic civilization were Muslim, many others were Christian and Jewish.

[2] I have adopted a simplified transliteration system of Arabic terms that does not include full vocalizations with diacritical marks.

(*topos*), but Plato was perhaps the first amongst them to systemically inquire about its *essence*. The Platonic reflections on the *quiddity* of spatiality were principally gathered in the dialogues of the *Timaeus* in reference to what is named by the Greek appellation *khôra* (*chora*), which is customarily translated in several modern European languages as: *space*, *espace*, or *Raum*. Nonetheless, the notion of spatiality, as that which is akin to *extension*, or to the isotropic and homogeneous conception of mathematical space, does not squarely correspond with what is intended by the signifier *khôra*; rather, translation involves in this regard some sort of semantic and representational transformation, while it also points to historical developments in the unfolding of the concept of space.

As it was ambiguously relegated to us by Plato (on the authority of the narrative of the Pythagorean astronomer, Timaeus of Locri), it is said that *khôra* is a "third genus" (*triton*), besides being and becoming, which is in itself neither intelligible nor sensible.[3] As a "receptacle", this "boundless" *khôra* receives all becoming entities without taking on the character of what it contains. It is therefore amorphous and characterless.[4] Moreover, like the forms (*eidoi*), it is everlasting and does not admit of destruction.[5] These ambivalent propositions concerning the reality of *khôra* may have indeed constituted the earliest systemic philosophical and metaphysically-oriented reflections on the nature of "spatiality" in the context of Ancient cosmology and classical physics.

Based on Aristotle's endeavor to define "place" (*topos*), it was reductively conjectured that Plato's *khôra* referred to prime matter. However, this exegesis served the purposes of the Aristotelian conception of "place" as a mode of containment by envelopment, more than that it resulted necessarily from a faithful and attentive reading of Plato's *Timaeus*. After all, Aristotle rejected the theories that posited *place* as being the form (*eidos*), the matter (*hulê*; partly following his own interpretation of Plato's *khôra*), or the interval (*diastêma*) between the extremities of the body that it contains.[6] He rather defined *topos* as "the innermost primary surface-boundary of the

[3] Plato, *Timaeus*, 48E, 52A-B. Translations for this paper from, Plato, *Timaeus*, trans. R. G. Bury, with parallel Greek text (Cambridge, MA: Harvard University Press, 1929), Loeb Classical Library 234, 8th repr. 1999. I have also examined this notion of the *khôra* (*chora*), elsewhere in: El-Bizri, Nader, "*On kai khôra*: Situating Heidegger between the *Sophist* and the *Timaeus*, *Studia Phaenomenologica*, IV, no. 1-2 (2004): 73-98; El-Bizri, Nader, "*Ontopoiêsis* and the Interpretation of Plato's Khôra," *Analecta Husserliana: The Yearbook of Phenomenological Research*, LXXXIII (2004): 25-45; El-Bizri, Nader, "*Qui-êtes vous Khôra?* Receiving Plato's *Timaeus*", *Existentia Meletai-Sophias*, XI, Issue 3-4 (2001): 473-490.

[4] Plato, *Timaeus*, 50B-51A.

[5] Plato, *Timaeus*, 52A-B.

[6] Aristotle, *Physics*, IV, 212a 3-5. For this paper, translations taken from Aristotle, *Physics*, ed. W. David Ross (Oxford: Oxford University Press, 1936).

containing body that is at rest, and is in contact with the outermost surface of the mobile body it contains."[7]

Even though Aristotle affirmed that *topos* has the three dimensions of length, width, and depth,[8] he nonetheless seemed to indicate in his conception of spatiality by way of "containment" that a *place* is ultimately a two-dimensional "surface-limit" of "envelopment". Furthermore, he distinguished between what may be called "a *local* place" (which consists of the *specific surfaces* of *the containing body* that a given thing occupies), and a contrasting "*cosmic natural place*", namely the one to which things tend to return, due to their own nature, if not prevented from doing so; like heavy bodies *by their nature travel downwards* to the Earth, in a fall in the direction of the center of the Universe, and light bodies *by their nature travel upwards* to the heavens.[9] This view accentuated also the Aristotelian presupposition of the existence of a certain "*power* (*dunamis*) of place", by way of asserting the existential anteriority of *topos* with respect to all beings.[10]

Mathematical Space

The most significant critique directed at the Aristotelian definition of *topos* was accomplished through the geometrical conception of place (*al-makan*) by the Arab polymath al-Hasan Ibn al-Haytham (known in Latin as Alhazen; b. 965 CE Basra, d. ca. 1041 CE, Cairo).[11] Using mathematical demonstrations, in reference to geometrical figures of equal surface-areas, based on studies conducted on figures of equal perimeters, Ibn al-Haytham demonstrated that the sphere is the largest in size with respect to all other solids with equal areas for their enveloping outer surfaces.[12]

In contesting the longstanding Aristotelian *physical* conception of *topos* as a boundary surface of "containment by way of envelopment", Ibn al-Haytham posited *al-makan* (place) as an "imagined [postulated] void" (*khala' mutakhayyal*) whose existence, as an invariable geometric entity, is secured in the imagination. He moreover held that the "postulated void" *qua* "mathematical place" consisted of imagined immaterial distances that are between the opposite points of the surfaces surrounding it. He furthermore noted that the imagined (mathematical) distances of a given body, and those of its containing place, get superposed and united in such a way that they

---

[7] Aristotle, *Physics*, IV, 212a 20-21.

[8] *ibid*, IV, 209a 5.

[9] *ibid*, IV, 4, 212a24.

[10] *ibid*, IV, 208b 33-34, 209a 1-2).

[11] Ibn al-Haytham's critical thesis was presented in his *Discourse on Place* (*Qawl fi al-makan*); for the Arabic critical edition and annotated French translation of this tract (*Traité sur le lieu*) see: Rashed, Roshdi, *Les mathématiques infinitésimales du IXe au XIe siècle*, Vol. IV (London: al-Furqan Islamic Heritage Foundation, 2002), pp. 666-685.

[12] Rashed, *op cit.*, Vol. I, p 776, p. 828; Vol. II, pp. 381-382, pp. 451-457.

become the same congruent distances (*qua* dimensions), namely as the magnitudes of mathematical lines having lengths without widths.

Ibn al-Haytham argued that his geometrical conception of place was ultimately neutral from the standpoint of "ontology", given that it was not simply obtained through a "theory of abstraction" as such, nor was it derived by way of a "doctrine of forms", nor was it grasped as being a phenomenal object of "immediate experience" or "common sense". It is rather the case that his *geometrized* place resulted from a mathematical isometric "bijection" function between two sets of relations/distances.[13] Nothing is thus retainable of the properties of a body other than its *extension*, which consists of mathematical distances that underlie the geometrical conception of the place it occupies. Accordingly, the *place* of a given object is a "*region of extension*" that is defined by the distances between its points, and on which the distances of that object can be applied and superposed.[14] This mathematical development found later affirmations in the history of science and philosophy in the conception of *place* as a *space*; namely, as it was later the case with Descartes' notion of "*extensio*" and Leibniz's "*analysis situs*".[15] A geometrical place is hence posited as a "metric" of a region of (the so-called) "Euclidean" *qua* "geometrical *space*", which is conceived extensionally. Consequently, the geometrization of place points to what was later embodied in the conception of the "anteriority of spatiality" over the demarcation of a metric of its regions by means of mathematical lines and points, as explicitly implied by the notion of "space".[16] After all, the concept of a homogeneous "Euclidean space" is a relatively modern invention that coincides with the development of the Renaissance *perspectivae* traditions that were influenced (among others) by Ibn al-Haytham's *Optics* (Arabic: *Kitab al-Manazir*; Latin: *De aspectibus* or *Perspectivae*), and that eventually led to the formation of the early-modern notion of a "Cartesian space". After all, Euclid noted in his *Data* Proposition 55 (related to his *Elements* VI, Proposition 25) that "if an

---

[13] "Bijection" designates an equivalence relation or function of mathematical transformation that describes a "one-to-one" correspondence (or "injection") and a "surjection" ("on-to") between two sets.

[14] Rashed, Vol. IV, *op cit.*, pp. 658, 901.

[15] Rashed, Vol. IV, *op. cit.*, pp. 661-662, and associated notes 25-26 on p. 662. It is also pertinent to note in this regard what Descartes stated, namely that: "*L'objet des géomètres, que je concevais comme un corps continu, ou un espace indéfiniment étendu en longueur, largeur et hauteur ou profondeur, divisible en diverses parties, qui pouvaient avoir diverses figures et grandeurs, et être mues ou transposées en toutes sortes*". See : Descartes, René, *Discours de la méthode*, in *Œuvres de Descartes*, eds. Charles Adam and Paul Tannery (Paris: Vrin, 1965), Vol. 6, p. 36 --also cited by Rashed, p. 662. Moreover, Leibniz noted that a place (*situs*) is a fragment of the geometrical *space* that describes an invariable relation between the points of a given configuration of an object, like [A•B] which designates an *extensum* that ties [A] with [B] mathematically with invariance. See: Leibniz, Gottfried Wilhelm, *La Caractéristique géométrique*, ed. Javier Echeverria, trans. Marc Parmentier, *Mathesis* series (Paris: Vrin, 1995), p. 235 --also quoted by Rashed, p. 662

[16] Rashed, vol. IV, *op. cit.*, pp. 661-662.

area (*khôrion*) be given in form and in magnitude, its sides will also be given in magnitude". The expression deployed by Euclid that is closest to a notion of space *qua khôra*, is "*khôrion*" as "an area enclosed within the perimeter of a specific geometric figure."[17]

Not uncommon in the mediaeval intellectual history in Islam, was the fact that selected problems in theoretical philosophy were solved, or attempted to be resolved, with the assistance of mathematics. This method is the one that Ibn al-Haytham adopted in demonstrating his geometrical definition of *al-makan* (place) as a solution to a longstanding problem that remained philosophically unresolved, which, to our knowledge, also constituted in its own right the first demonstrated attempt to geometrize "place" in the history of mathematics and science. Ibn al-Haytham's primary objectives aimed at promoting a geometrical conception of place that is akin to *extension* in an attempt to address selected mathematical problems that emerged in reference to unprecedented developments in geometrical transformations (*naql*), the introduction of motion (*haraka*) in geometry, the anaclastic research in conics and dioptrics in the ninth/tenth century prolongations of the Apollonian-Archimedean Arabic school in mathematics.[18]

Besides the penchant to offer mathematical solutions to problems in theoretical philosophy, which were challenged by longstanding historical obstacles and epistemic impasses, Ibn al-Haytham's remarkable and successful endeavor in geometrizing place was undertaken in view of sustaining and grounding his research in mathematical analysis and synthesis (*al-tahlil wa-al-tarkib*),[19] and in response to the needs associated with the unfurling of his studies on "knowable [mathematical] entities" (*al-ma'lumat*)

---

[17] Euclid, *The Thirteen Books of Euclid's Elements*, vols. 1-3, translated with introduction and commentary by Thomas L. Heath (New York: Dover Publications, 1956). I also refer the reader to: Vesely, Dalibor, *Architecture in the Age of Divided Representation: The Question of Creativity in the Shadow of Production* (Cambridge, Mass.: MIT Press, 2004), pp. 113, 140-141; Lachterman, D.R., *The Ethics of Geometry: A Genealogy of Modernity* (London: Routledge, 1989), p. 80; Kline, M., *Mathematics: The Loss of Certainty* (Cambridge: Cambridge University Press, 1980), p. 87. I have also examined this question elsewhere in the following studies: El-Bizri, Nader, "In Defence of the Sovereignty of Philosophy: al-Baghdadi's Critique of Ibn al-Haytham's Geometrisation of Place", *Arabic Sciences and Philosophy* (Cambridge University Press), 17, Issue 1 (2007): 57-80; El-Bizri, Nader, "A Philosophical Perspective on Alhazen's *Optics*", *Arabic Sciences and Philosophy*, 15, Issue 2 (2005):189-218 (Cambridge University Press); El-Bizri, Nader, "*La perception de la profondeur: Alhazen, Berkeley et Merleau-Ponty*", *Oriens-Occidens: sciences, mathématiques et philosophie de l'antiquité à l'âge classique* (*Cahiers du Centre d'Histoire des Sciences et des Philosophies Arabes et Médiévales, CNRS*), Vol. 5 (2004):171-184.

[18] Namely, the legacy of mathematicians like the Banu Musa ibn Shakir (The sons of Musa Ibn Shakir), Thabit ibn Qurra, Ibrahim ibn Sinan, Abu Sa'd al-'Ala' ibn Sahl, Abu Sahl Wayjan ibn Rustam al-Quhi, and Ahmad ibn Muhammad ibn 'Abd al-Jalil al-Sijzi.

[19] The Arabic critical edition and the annotated French translation of this treatise (*Fi al-tahlil wa-al-tarkib*; *L'Analyse et la synthèse*) are established in Rashed, *op. cit.*, Vol. 4 (2002), pp. 230-391.

in order to reorganize most of the notions of geometry and rethinking them in terms of motion.[20] Consequently, he had to critically reassess the dominant philosophical conceptions of place in his age, which were encumbered by inconclusive theoretical disputes that were principally developed in reaction to Aristotle's *Physics*.

One ought to add here that, while most philosophers adopted the Aristotelian conception of place (including Ibn Sina [Avicenna] in *Kitab al-Shifa'* and *Kitab al-Hudud*,[21] respectively *The Book of Healing* and *The Book of Definitions*), the dialectical theologians (mainly the exponents of *Mu'tazilite kalam*) affirmed the existence of the void, and reflected on place as being akin to spatiality (*hayyiz* or *tahayyuz*) in deliberations that were partly founded on geometric adaptations of the physical theories of Greek atomism.

Physical Place

Aristotle's definition of place received bold classical critiques in the commentaries on his work, including the reflections of Theophrastus on this matter and the poignant objections advanced by Philoponus in support of a conception of *topos* as extension or interval (*diastasis*; *diastêma*). Additional doubts concerning Aristotle's conception of *topos* were also delineated in Simplicius' corollary on place.[22] However, what primarily distinguishes Ibn al-Haytham from his predecessors is that his critique of Aristotle was mathematical, and, that it was partly auxiliary to his response to the epistemic and mathematical needs to geometrize place, while what preceded his efforts (including Philoponus' corollaries) mainly restricted their critical objections to the Aristotelian notion of *topos* to philosophical deliberations in classical physics.

Critical objections were leveled at Ibn al-Haytham's geometrization of place by philosophers of the Aristotelian tradition in mediaeval Islamic civilization; similarly as was the case with 'Abd al-Latif al-Baghdadi (d. ca. 1231 CE) in his tract *Fi al-radd 'ala Ibn al-Haytham fi al-makan* (*A*

---

[20] The Arabic critical edition and annotated French translation of this treatise (*Fi al-ma'lumat*; *Les connus*) are established in Rashed, *op. cit.*, Vol. IV (2002), pp. 444-583. See also: Rashed, Roshdi, "La philosophie mathématique d'Ibn al-Haytham, II: *Les Connus*," *MIDEO*, 21 (1993): 87-275.

[21] Ibn Sina, *Kitab al-Hudud*, ed. A.-M. Goichon (Cairo: Institut Français d'Archéologie Orientale du Caire, 1963).

[22] I refer the reader to: *Simplicii in Aristotelis Physicorum Libros Quattuor Priores Commentaria*, ed. H. Diels, *Commentaria in Aristotelem Graeca*, Vol. IX (Berlin, 1882); Simplicius, *Corollaries on Place and Time*, trans. J. O. Urmson (London: Duckworth, 1992), pp. 601-611. See also: Simplicius, *On Aristotle, Physics 4.1-5, 10-14*, trans. J. O. Urmson (London: Duckworth, 1992); Philoponus, *Corollaries on Place and Void*, and: Simplicius, *Against Philoponus on the Eternity of the World*, trans. D. Furley and C. Wildberg (London: Duckworth, 1991).

refutation of Ibn al-Haytham's place).[23] Closely following each of Ibn al-Haytham's arguments, and not failing to admire the mathematical acumen of the author subjected to his critique, al-Baghdadi claimed that Ibn al-Haytham did not logically account for a correspondence/concomitance between a given object and its "place" *qua* "enveloping surfaces" (*sath muhit*) as both being subject to change.[24]

According to al-Baghdadi, Ibn al-Haytham's geometrical proofs neglected the fact that a change in a given object leads to a transformation in its shape, the total sum of its surface areas, and the place it occupies. Al-Baghdadi presupposed philosophical accounts of the individuation of bodies as a modality by virtue of which he attempted to offer counterexamples to Ibn al-Haytham's geometrical demonstrations. For instance, al-Baghdadi argued that the judgment of a given body in-itself differs from judging its surrounding surfaces; since the surfaces of a body change in the magnitude of their areas with the transformation of the shape of that body, while the body is unchanged in-itself.[25] He thus believed that Ibn al-Haytham's mathematical doubts were not only raised with respect to place as an enveloping surface, but were moreover applicable to the essence of the body that occupies it; given that a body is in a place by way of its actual surfaces not its internal potential distances. He was also unsure whether Ibn al-Haytham considered the distances of a body and those of its place as being potentialities and not actualities; hence, positing them as non-existents.[26] Al-Baghdadi asserted that the mathematician judges distances insofar that they are imagined in the mind as being abstracted from matter, while the physicist grasps them as existing externally. His critique was principally guided by Aristotelian metaphysical concerns, and it ultimately failed to recognize the epistemological significance of Ibn al-Haytham's mathematical definition of place and its ontological neutrality.

Theological Accounts of Place and Space

Reflections on the nature of place/space in the history of ideas in Islam in the classical period were not restricted to the domains of philosophy or the sciences; they rather carried significant theological implications, particularly when accounted for in terms of meditations on the question concerning the divine essence and attributes.[27] For instance, the celebrated metaphysician

[23] The Arabic edition and annotated French translation of this treatise (*Fi al-radd 'ala Ibn al-Haytham fi al-makan*; *La réfutation du lieu d'Ibn al-Haytham*) are established in Rashed, *Les mathématiques infinitésimales*, *op. cit.*, Vol. IV, pp. 908-953.

[24] Rashed, Vol. IV, *op. cit.* pp. 914-915.

[25] *ibid,* Vol. IV, pp. 924-925.

[26] *ibid,* Vol. IV , pp. 916-917.

[27] I have addressed some of the theological aspects of this question elsewhere; see: El-Bizri, Nader, "God: essence and attributes", in Winter, Tim (ed.), *The Cambridge Companion to Classical Islamic Theology*, (Cambridge: Cambridge University Press, 2008), pp. 121-140.

Ibn Sina (Avicenna) held that the "Necessary-Existent-due-to-Itself" (*wajib al-wujud bi-dhatihi*); namely, what in an ontological inquiry points to Divinity, has no genus (*jins*), nor a definition (*hadd*), nor a counterpart (*nadd*), nor an opposite (*did*), and is detached (*bari'*) from matter (*madda*), quality (*kayf*), quantity (*kamm*), place (*ayn*), situation (*wad'*), and time (*waqt*). [28] Moreover, philosophically-oriented exponents of dialectical theology (*kalam*) argued that the Divine does not occupy a given place/space, nor is He in time. For instance, the theologian al-Sharif al-Jurjani held in his *Sharh al-mawaqif fi 'ilm al-kalam* (*Commentary on the Principles of Dialectical Theology*)[29] that God is not in any spatial location (*jiha*), or in a place (*makan*), unlike what was claimed by the exponents of anthropomorphism (*al-mushabbiha*; namely those who assign anthropomorphic or anthropocentric attributes to the divine in literal readings of scripture). Moreover, the *Ash'arite* theologian al-Amidi argued in his *Ghayat al-maram fi 'ilm al-kalam* (*The Principal Objectives of Dialectical Theology*) that God is not in a given place that contains Him nor is He in time.

Dialectical theologians attempted to show that if God were to be in a place, then the eternity of that spatial location would have been necessarily implied. However, they also argued that they have demonstrated with evidence that there is no eternal being but God, and that this constitutes a matter that is sustained through unanimous consensus (*ijma'*) amongst Muslims. They also held that what occupies a given place (*mutamakkin*) requires its own specific *situs* in such a way that its own existence is impossible without it; since they argued that the space-occupant (*mutamakkin*) is the *jawhar* (*substance*; though grasped by them as *atom* rather than the Aristotelian *ousia* [*substantia* or *essentia*]). Yet, a place dispenses of what occupies it, since it is possible to have a void, and this necessitates the emplacement (place-occupation) of the Necessary Being (*al-wajib*; namely what ontologically designates "God"), as well as the necessitation of place, and both are theologically considered to be false propositions. If the Necessary Being is in a place, then He will require His own place in such a way that His existence is impossible without it. Yet, this state of affairs does not hold since the Necessary Being cannot but necessarily exist as what is

---

[28] Ibn Sina, *Kitab al-Hidaya*, ed. Muhammad 'Abdu (Cairo, 1874), pp. 262-263; Ibn Sina, *Kitab al-Shifa', Metaphysics II*, (eds.) G. C. Anawati, Ibrahim Madkour, Sa'id Zayed (Cairo, 1975), p. 354; Salem Mashran, *al-Janib al-ilahi 'ind Ibn Sina* (Damascus, 1992), p. 99; I have also investigated this question elsewhere in the following studies: El-Bizri, Nader, "Avicenna and Essentialism," *Review of Metaphysics*, Vol. 54 (2001): 753-778; El-Bizri, Nader, "Being and Necessity: A Phenomenological Investigation of Avicenna's Metaphysics and Cosmology," in Tymieniecka, Anna-Teresa (ed.), *Islamic Philosophy and Occidental Phenomenology on the Perennial Issue of Microcosm and Macrocosm* (Dordrecht: Kluwer Academic Publishers, 2006), pp. 243-261.

[29] Al-Sharif al-Jurjani, *Sharh al-mawaqif fi 'ilm al-kalam*, ed. Ahmad al-Mahdi (Cairo: *Maktabat al-qahira*, 1976), Part V, Section 2.

self-subsistent. One could then argue that the emplacement of the Necessary Being is either a necessity or that this Necessary Being does not need to be in a place. If we say that the Necessary Being has to be necessarily in a place and cannot be otherwise, then we cannot still claim that in this case a place still dispenses of what is in it. Accordingly, the necessitation of the emplacement of the Necessary Being, and the necessitation of place, both were theologically taken to be false propositions.

The exponents of *kalam* grasped place as a void (*khala'*), which is an existing dimension that does not subsist in matter.[30] The void is a dimension that has been created, and yet that does not exist in the same way as embodied beings exist. The *kalam* conception of space as *hayyiz* refers to spatiality as a phenomenon of spacing, namely as the apportioning of a place that is occupied by an atom. After all, the physics of the exponents of *kalam* was principally inspired by Greek atomism,[31] and this arguably facilitated their attempted rejection of the views of the philosophers (*falasifa*; *hukama'*) who were primarily influenced by Aristotelian physics. However, a theologian like al-Amidi argued in his *Sharh alfaz al-hukama'* (*Commentary on the Lexicon of the Philosophers*) that a *hayyiz* (*qua space*) is the apportioning/measuring of place (*taqdir al-makan*), and yet that place (*makan*) is the inner surface of the containing body that is in contact with the outer surface of the contained body. Based on this, he seems to combine the classical *kalam* physical theory (which takes space/place to be the portion in the void that is occupied by an atom) with the Aristotelian definition of *topos*.

However, it was a distinctive aspect of later *kalam* schools that they integrated elements of Peripatetic philosophy into their theological systems, including the joining of incompatible physical theories (atomist versus Aristotelian) along with their entailed anomalies. For instance, al-Jurjani rectifies al-Amidi's definition by re-asserting the atomist thesis that place (*makan*) is the imagined void that a body occupies; coming in this case closer to the definition of place by Ibn al-Haytham. Al-Jurjani elaborates on this point by stating that the imaginary void that is occupied by an extended thing, or a void that is occupied by that which is un-extended (like an atom), would itself be un-extended, and yet still having a given magnitude. If God was space-occupying (*mutahayyiz*), then He would have been equated with all the space-occupants in *quiddity*. However, this necessitates either the eternity of bodies or that He has been created (*muhdath*); since, equivalents agree in properties. Reflections on place in terms of accounts pertaining to the divine essence and attributes resulted in theological difficulties when faced with the interpretation of Qur'anic verses like: "The All-Compassionate sat Himself on the Throne" (Qur'an 20:5); or: "To Him ascends the good word" (Qur'an

[30] Al-Amidi, *Sharh alfaz al-hukama'* (*Commentary on the Lexicon of the Philosophers*), ed. 'Abd al-Amir al-A'sam (Beirut: *Dar al-manahil*, 1987), p. 86.
[31] Dhanani, Alnoor, *The Physical Theory of Kalam* (Leiden: Brill, 1994).

35:10); or: "He is the Lord of the ascents, by which the angels and the spirit mount up to Him" (Qur'an 70:4). For instance, the theologian Hasan Jalabi al-Fanari (fl. 15th cent.) held that "sitting firmly on the Divine Throne" implies space-occupation (*tahayyuz*). Moreover, the image of *ascent* suggests a motion upwards; or at least a movement from Earth towards the Heavens. Furthermore, pointing to the sky in worship or in the invocation of God is also another way by which one is sometimes misled into believing that a sense of space-occupation is entailed by it. Based on Jalabi's scriptural exegesis and hermeneutics, pointing to the sky does not literally mean that the Lord is in the sky as a given spatial region, rather that this pointing posits a certain orientation which is akin to the *Qibla* that demarcates the direction of Mecca for the worshippers.[32] However, the theologians who advocated the views of the commanding jurist Ahmad Ibn Hanbal (8th-9th century) asserted that "sitting on the Throne" is a known matter, while the modality by virtue of which it happens is unknowable, and that, ultimately, inquiries about this are at best heterodox and in some cases heretic.

Time: Classical Conceptions of Time

As with the case of philosophical reflections on the notions of space and spatiality, classical conceptions of time and of temporality confronted philosophers with uneasy paradoxes. Some wondered whether time was altogether nonexistent, while others doubted the reality of its divisibility into parts by arguing that the *past* ceased to be, that the *future* does not yet exist, and that the *present* as a moment/now; that is, without magnitude, and thus is not a real part of time. Additionally unclear in these debates, was whether the passage of time progressed with smooth continuities, or whether it proceeded by way of discontinuous and divisible leaps. Even though the inquiries about the nature of time were essentially integrated within classical physical theories about motion (*kinêsis* and change *qua metabolê*), their cosmological and metaphysical bearings subsequently impacted the unfurling of philosophical and theological mediaeval speculations about creation and causation.

In Plato's *Timaeus*,[33] time (*khronos*) was grasped as a moving image (*eikona*) that came into existence with the generation of the heavens, and which imitated eternity by circling round. In the earliest systemic philosophical investigation of the essence and existence of time, which was contained in Aristotle's *Physics*,[34] *khronos* was defined as the [measuring] number (*metron*) of a continuous motion (*kinêsis*) with respect to the anterior (*proteron*) and the posterior (*husteron*). Rejecting the claim that time was the

[32] Jalabi's views are incorporated as commentaries in the notes apparatus of the Arabic edition of al-Jurjani's *Sharh al-mawaqif, op. cit.*; note 3, pp. 37-38.
[33] Plato, *Timaeus, op. cit.*, 37d-38a.
[34] Aristotle, *Physics, op. cit.*, 219b3-4; 220a25-b20; 222b20-23.

movement of "the whole", Aristotle argued that the circular, uniform, and continuous motion of the celestial sphere (*sphaira*) acted as a measure (*metron*) of time.[35] Moreover, the Aristotelian conception of *khronos* had affinities with the notion of *ekstatikos*, as the mode of undoing beings, which is implied by the processes of motion that entailed change *qua metabolê*.[36]

Aristotle's theory subsequently received numerous responses by Neo-Platonist commentators and Hellenist exegetes (as principally grouped in the monumental: *Commentaria in Aristotelem Graeca*).[37] For instance, Damascius argued that time was a simultaneous whole, while Plotinus grounded its reality on the changing life of the soul.[38] As for Simplicius, he defended the thesis of the eternity of the world against doubts raised by the grammarian John Philoponus, who arguably adopted a Christian doctrine of *creatio ex nihilo*. Moreover, the author of the *Confessiones*,[39] Augustine of Hippo, noted that *tempus* (time) was created when the world came to be, while affirming that the existential reality of time is grounded in the present (*praesens*), which in itself is what tends not to be (*tendit non esse*), given that only eternity was stable (*semper stans*). On his view, temporality is also marked by *distensio*, namely dilatation or extension.[40] Based on a belief in the linear directionality of time, from Genesis to Judgment, Augustine argued that the *presence* of *past* things was preserved in memory, the *presence* of manifest (*present*) things was confirmed by perception, and that the *presence* of things *future* was highlighted by expectation. Accordingly, the reality of time depended on an *anima* (soul) that remembers, perceives, and anticipates events; partly echoing in this Aristotle's claim in the *Physics*[41] that *khronos* required a soul or an intellect (*psukhês nous*) to number it (*arithmein*).[42]

Time in Mathematics and the Exact Sciences

The reception and adaptive assimilation of the Greek conceptions of time by scholars in mediaeval Islamic civilization, varied in terms of the levels of adherence to the sources, and in terms of the reformative aspects of associated commentaries or conceptual prolongations in rethinking these notions. While philosophers (*al-hukama'*; *al-falasifa*) of the Peripatetic and

---

[35] *ibid.*, 223b21.
[36] *ibid.*, 222b.
[37] Philoponus, *Corollaries on Place and void, op. cit.*; Simplicius, *Corollaries on Place and Time, op. cit.*; also see Sorabji, Richard, *Time, Creation and the Continuum* (Ithaca: Cornell University Press, 1983).
[38] *Enneads*, 3.7.11-13. translations from Plotinus, *Enneads*, trans. Arthur Hilary Armstrong, with parallel Greek text (Cambridge, Mass.: Harvard University Press, 1966-1967).
[39] Augustine, *Confessions*, ed. James O'Donnell (Oxford: Clarendon Press, 1992).
[40] *ibid.*, XI, sect. 23.
[41] Aristotle, *Physics, op. cit.*, 218b29-219a1-6, 223a25.
[42] I have investigated related topics in: El-Bizri, Nader, "Avicenna's *De Anima* between Aristotle and Husserl," in Tymieniecka, Anna-Teresa (ed.), *The Passions of the Soul in the Metamorphosis of Becoming*, (Dordrecht: Kluwer Academic Publishers, 2003), pp. 67-89.

Platonist traditions tended to find innovative extensions of the views of the Ancients within monotheistic outlooks on time, the dialectical theologians (*al-mutakallimun*; i.e. the exponents of *kalam*) tended in general to object to some of the bearings of these "pagan" doctrines, and consequently developed novel ontological-theological accounts regarding eternity, perpetuity, and temporality.[43] However, the conception of time and the techniques deployed in its measurement within the history of ideas in Islam were not restricted to the doctrines of the philosophers or the theologians; rather, accomplished investigations in this regard were also conducted in classical traditions in science and mathematics that built on the legacies of the likes of Euclid, Archimedes, Ptolemy, Apollonius of Perga, and Heron of Alexandria, as well as referring to Plato and Aristotle.

The research in geometry, arithmetic, algebra, astronomy, optics and mechanics in mediaeval Islamic civilization (principally: 9th-14th century CE),[44] offered solid foundations for the design, construction, and perfection of time-measurement devices and instruments, including tools like astrolabes,[45] sundials, water-clocks (*cum* automata) and compasses. These investigations assisted also in devising the theoretical and geometrical models for the design of optical tools in the sciences of catoptrics and dioptrics (respectively: the science of the reflection of light and the science of the refraction of light, with their related instruments).[46] Such models and

---

[43] I have discussed this topic in length elsewhere; see: El-Bizri, Nader, "Some Phenomenological and Classical Corollaries on Time," in Tymieniecka, Anna-Teresa (ed.), *Timing and Temporality in Islamic Philosophy and Phenomenology of Life* (Dordrecht: Kluwer Academic Publishers, 2007), pp. 137-155; El-Bizri, Nader, "Time (Concepts)," in Meri, Josef W. (ed.), *Medieval Islamic Civilization: An Encyclopedia*, (New York, London: Routledge, 2005), Vol. II, pp. 810-812.

[44] One could mention here the polymaths: Muhammad Ibn Musa al-Khwarizmi (d. 850 CE), the Banu Musa (Sons of Musa Ibn Shakir; fl. 9th century CE, Baghdad), Ya'qub Ibn Ishaq al-Kindi (d. 873 CE), Thabit Ibn Qurra (d. 901 CE), Abu 'Abd'Allah al-Battani (Albategnius; d. 929 CE), Ibrahim Ibn Sinan (d. 946 CE), Abu Sa'd al-'Ala' Ibn Sahl (d. 1000 CE), Abu Sahl Wayjan Ibn Rustam al-Quhi (d. 1000 CE), Ahmad Ibn Muhammad Ibn 'Abd al-Jalil al-Sijzi (d. 1020 CE), Abu 'Ali Ibn Sina (Avicenna; d. 1037 CE), al-Hasan Ibn al-Haytham (Alhazen; d. ca. 1041 CE), Abu Rayhan al-Biruni (d. 1048 CE), 'Umar al-Khayyam (d. ca. 1129 CE), Ibn al-Razzaz al-Jazari (fl. 13th century CE), Nasir al-Din al-Tusi (d. 1274 CE), and Kamal al-Din al-Farisi (d. 1320 CE).

[45] Astrolabes could not have been perfected unless greater accomplishments have been made in the domain of spherical geometry, given that these instruments presupposed a careful and accurate projection of the forms that are mathematically postulated as being on curved-spherical surfaces unto rectilinear planar surfaces.

[46] For instance, 10th century research on anaclastic (refractive) curves as sections in conics (parabola, hyperbola, ellipse, convex and bi-convex curves) offered geometrical models for optical studies in catoptrics and dioptrics in view of perfecting lenses, as manifest in the works of Ibn Sahl, al-Quhi, and al-Sijzi, with extensions of their findings in the investigations of Ibn al-Haytham and Kamal al-Din al-Farisi. This mathematical research involved the introduction of motion in geometry, and the use of geometrical transformations, not only in reference to figures, but also to their spatial relations.

devices were of great value for later developments in the observations of astronomy in reference to the motions of the heavenly spheres and their cycles. The applications of this research assisted also the science of timekeeping (*'ilm al-mawaqit*) and the establishment of calendars (*taqawim*) to serve particular religious purposes,[47] or to furthermore support studies in meteorology and the concrete determination of timing in navigation.

Ishaq Ibn Hunayn's (fl. 9th century CE; Baghdad) translation of Aristotle's *Physics* (*al-Tabi'a*) acted as a principal source for the transmission of the Aristotelian conception of *khronos* (*al-zaman*) into Arabic, which subsequently inspired variegated emergent philosophical interpretations of time in the history of ideas in Islam. For instance, al-Kindi (d. ca. 873 CE) held that *al-zaman* (time) had a beginning and an end, and that it measured motion according to number, while al-Farabi (Alfarabius; d. 950 CE) and Ikhwan al-Safa' (The Brethren of Purity; fl. 10th century CE, Iraq)[48] affirmed that time resulted from the movement of the celestial sphere (*al-falak*). As for Abu Bakr al-Razi (Rahzes; d. 930 CE), he claimed that the perpetuity (*dahr*) was absolute (*mutlaq*), while construing time (*al-zaman*) as being a flowing substance (*jawhar yajri*) that is bound (*mahsur*) as well as being associated with the motion of *al-falak*.

In *Kitab al-Hudud* (*The Book of Definitions*), Ibn Sina (Avicenna) defined *al-zaman* (time) as that which imitates the created being (*yudahi al-masnu'*), or is in its image, and acts as the measure of motion (*miqdar al-haraka*) in terms of the anterior and the posterior (*al-mutaqaddim wa-al-muta'akhkhir*). He also noted that supra-temporal duration (*al-dahr*) resembled the Creator (*yudahi al-sani'*) insofar that it was stable throughout the entirety of time. Moreover, in the *Kitab al-Isharat wa-al-tanbihat* (*Book of Pointers and Directives*),[49] Ibn Sina linked time to physical inquiries about motion, and in *'Uyun al-hikma* (*Essences of Wisdom*) he construed it as a quantity (*kammiyya*) of motion that measures (*yuqaddir*) change, and whose perpetuity (*dahr al-haraka*) generated temporality.

---

[47] Time measurement (*tawqit*) is central to the determination with accuracy of the timings of the decreed five daily prayers in Islam, and in supporting the observations in astronomy for demarcating the beginning of the fasting month of *Ramadan* and its ending with the start of *'Id al-fitr*, which depend on a developed coordinative system to compute time in the lunar cycle, with its temporal shifts with respect to the solar calendar and seasons. For a study on some applications of the science of timekeeping in Islam, see Kennedy, E. S., "Al-Biruni on the Muslim Times of Prayer," in Chelkowski, Peter J. (ed.), *The Scholar and the Saint: Studies in the Commemoration of Abu'l-Rayhan al-Biruni and Jalal al-Din al-Rumi*, (New York: New York University Press, 1975), pp. 83-94.

[48] Ikhwan al-Safa', *Rasa'il Ikhwan al-Safa'*, ed. Butrus Bustani (Beirut: *Dar Sadir*, 1957).

[49] Ibn Sina, *Kitab al-Isharat wa-al-tanbihat*, ed. Sulayman Dunya (Cairo: *Dar al-ma'arif bi-misr*, 1957-1960).

Time played also a notable role in Ibn al-Haytham's *Kitab al-Manazir* (*The Optics*)[50] wherein it was shown through experimental means (*i'tibar*) that the propagation of light rays was subject to time. Ibn al-Haytham consequently inferred that the velocity of light (*al-daw'*) was finite despite being immense in magnitude. He moreover held that acts of visual discernment and comparative measure (*al-tamyiz wa-al-qiyas*), which constitute some of the principal psychological-physiological determinants of vision, were subject to the passage of time even if not felt by the beholder. He also cautioned that: if the temporal duration of contemplative or immediate visual perception fell outside a moderate range it then resulted in optical errors. In addition, he listed *al-zaman* as one of the known entities (*ma'lumat*) while taking duration (*mudda*) to be its essence (*mahiyya*) and the scale (*miqyas*) measuring its magnitude (*miqdar*) and quantity, which become knowable by way of the observational methods of the science of astronomy in reference to the motion of the celestial sphere (*al-falak*).

Moreover, the 10th century mathematician Abu Sahl Wayjan Ibn Rustam al-Quhi (d. ca. 1000 CE) sought to geometrically establish the possibility of an infinite motion in a finite time (*fi al-zaman al-mutanahi haraka ghayr mutanahiya*); opposing in this the philosophical *communis opinio* of his age, which followed the doctrine advanced in Aristotle's *Physics.*[51] Accordingly, al-Quhi showed that if the arc of a given semicircle can be traversed in a finite time, its projected motion on an infinite branch of a hyperbola, which tends to infinity, is likewise covered in a finite time. His demonstration appealed to optics in postulating that the propagation of light in this projection was instantaneous; hence that the motion on the arc of the semicircle and that on the branch of the hyperbola were simultaneous; while taking the former to being uniform and considering the latter as being variable and unbound in its accelerating speed along the infinity of the hyperbolic curve.[52]

Time in Philosophy and Theology

Opposing the views of the Peripatetic (*masha'i*) philosophers in Islam, the exponents of *kalam* (dialectical theology) articulated alternative conceptions of time that rested on physical theories inspired by adaptations of Greek atomism.[53] Time was grasped by the *mutakallimun* (the dialectical theologians; mainly the *Mu'tazilites* of Basra in Iraq) as being a purported (*mawhum*; virtual) phenomenon of changing appearances and renewed

---

[50] Ibn al-Haytham, *Kitab al-Manazir*, ed. Abdelhamid I. Sabra (Kuwait: National Council for Culture, Arts and Letters, 1983); Ibn al-Haytham, *The Optics, Books I-III On Direct Vision*, trans. Abdelhamid I. Sabra (London: Warburg Institute, 1989).

[51] Aristotle, *Physics, op. cit.*; Book VI, 7, 238a20-37.

[52] Rashed, Roshdi, *Geometry and Dioptrics in Classical Islam* (London: al-Furqan Islamic Heritage Foundation, 2005), p. 986.

[53] Dhanani, *The Physical Theory of Kalam*, *op. cit.*

atomic events (*mutajaddidat*), whereby a discrete moment (*waqt*) replaced the concept of a continuous *zaman*. For instance, Ibn Mattawayh (a disciple of the *Mu'tazilite* Chief *Qadi* of Rayy: 'Abd al-Jabbar; both fl. 10th-11th century CE) held in his *Tadhkira fi ahkam al-jawahir wa-al-a'rad* (*Treatise on Substances and Accidents*) that accidents (*al-a'rad*) do not inhere in substances (*al-jawahir*; namely the "atoms") for even a moment (*la yujab lubuthuha abadan*), given that God recreates the world continually. Motivated by the early-*kalam* physical theory, though resisting its thrust, al-Nazzam (Ibrahim Ibn al-Sayyar; d. 845 CE) believed in the divisibility of particles *ad infinitum*, which entailed that a spatial distance with infinitely divisible parts requires an infinite time to be crossed unless its traversal proceeded by way of leaps (*tafarat*), echoing in this the Stoic views regarding the Greek notion of *halma* (leap).

Furthermore, as we have noted in reference to ontological and theological reflections on space and place in the context of accounting for the question concerning the divine essence and attributes, it is also the case that the "Necessary-Existent-due-to-Itself" (*wajib al-wujud bi-dhatihi*); namely, what in an ontological investigation points to the Divine, is not temporal, in the sense of being indeterminable against the horizon of temporality or time. Moreover, as the Ash'arite theologian al-Amidi argued in his *Ghayat al-maram fi 'ilm al-kalam* (*The Principal Objectives of Dialectical Theology*), God is not in time.

In doubting the doctrine of the eternity of the world in *Tahafut al-falasifa* (*The Incoherence of the Philosophers*),[54] Abu Hamid al-Ghazali (d. 1111 CE) attempted to show that duration (*mudda*) and time (*zaman*) were both created. Furthermore, he argued that "the connection between what is habitually (*bi-al-'ada*) taken to be a cause and what is customarily taken to be an effect was not necessary,"[55] given that observation only shows that they were concomitant/concurrent. Consequently, he proclaimed that the ordering relation of an antecedent cause with a consequent effect does not necessarily rest on an irreversible directionality in time. In defense of causation, Ibn Rushd (Averroes) argued in *Tahafut al-tahafut* (*The Incoherence of the Incoherence*)[56] that al-Ghazali's "refutation of the causal principle" entailed an outright rejection of reason (*'aql*), while asserting that the eternal (*al-qadim*) was timeless and that the world was subject to the workings of a continuous *zaman*. However, Ibn Rushd may have misinterpreted al-Ghazali's thesis by mistaking "the rejection of a necessary connection

---

[54] Abu Hamid al-Ghazali, *Tahafut al-falasifa (The Incoherence of the Philosophers)*, trans. Michael Marmura, with parallel Arabic text (Provo, Utah: Brigham Young University Press, 1997).

[55] *ibid*, p.166.

[56] Ibn Rushd, *Tahafut al-tahafut*, ed. Muhammad 'Abid al-Jabiri (Beirut: *Markaz dirasat al-wihda al-'arabiyya*, 1998).

between what is habitually taken to be a cause and its effect" with a "refutation of causation" outright. After all, al-Ghazali's doubts regarding the "necessary connection between cause and effect" reflected his belief in the existence a "contingent" sense of causation that embodied an inherent "habitual" course of nature, with which corresponded a deeply entrenched "custom" of knowing natural phenomena through *seeming* causal connections, in reflection of an "ordained" pattern as willed by the Divine. Hence, al-Ghazali's causation is "habitual" rather than "necessary", and this does not readily entail a refutation of the causal principle as much as showing its contingent character, wherein it is *believed* that Divine Volition (*irada*) breaks the habitual course of nature (and of causation) under exceptional circumstances; known in religious terms as "miracles" (like when Abraham was thrown in the fire and did not burn; Qur'an [21:69]: "O Fire! Be thou coolness and peace on Abraham"). Furthermore, in affirming the truth of Genesis, Moses Maimonides (Musa Ibn Maymun; d. 1204 CE) asserted in *Dalalat al-ha'irin* (*The Guide for the Perplexed*)[57] that time was created, given that the celestial sphere (*al-falak*) and its motion on which it depended were both generated.

Decidedly, in conclusion, although speculations about time continued with scholars of the caliber of the theologian Fakhr al-Din al-Razi (d. 1209 CE), the Sufi master Ibn 'Arabi (d. 1240 CE), the metaphysicians Mir Damad (d. 1631 CE) and Mulla Sadra (d. 1640 CE), the elucidation of the uncanny reality of time remained inconclusive, and its quotidian familiarity perplexingly enigmatic.

---

[57] Maimonides, Moses, *Dalalat al-ha'irin* (*The Guide for the Perplexed*), trans. M. Friedlander (New York: Dover, 1956).

Agency and Space in Darwin's Concept of Variation

Chunglin Kwa

In Kwa's discussion of evolutionary time, he positions Darwin's "natural selection" as heir to an 18th Century mechanicist understanding of nature's history, even when taking to account the important innovations that he brought to it. The key to a modern historical understanding of nature was the concept of variation. Variation was predicated on an understanding of the active role of organisms *vis-a-vis* their environment. When we place evolutionary theory in the larger framework of the new historical understanding of the 19th century, different conceptualizations of time and space come into focus. Time would become the relational sequence between events, rather than the neutral background on which timeless laws of nature unfold. Space became the theatre on which organisms act out the advantages which variations in their biological constitution had given them.

Darwin's conceptions of "natural selection" inherited certain early modern mechanist views of nature. This historical debt will become apparent when examining Bonnet's and Buffon's "evolutionary theories". In addition, a theoretical prerequisite to Darwin's variation can be found in certain aspects of the work of Kant, and these are examined in some detail. Darwin thought that his method of collecting facts in support of the evolutionary theory was "Baconian". His stated aim was to reconstruct the process of species formation using the "inductive" method of amassing factual evidence and persuasive metaphors. Certainties Darwin could not offer, but trustworthy probabilities, by reasoning his way to plausible outcomes, he could.[1] Darwin was indeed criticized by some for not conforming to the logic of the deductive scientific method.[2] While Darwin could not have cared less, his method was neither a form of inductive generalizing nor was it abduction, or inference to the best available hypothesis. Instead, Darwin's evolutionary theory was a special case of the new historical reasoning of the 19th century.

Evolutionary theory is, therefore, not a "theory" like in the theoretical sciences. As a historical framework, it encompasses several theories which also constitute fields of their own. Several concepts of evolutionary theory, notably variation and selection, are the well-known key concepts in Darwin's theory of evolution. One reason why it is difficult to evaluate them is that they function in several ways at the same time: in the

---

[1] Browne, Janet, *Charles Darwin: The Power of Place. Vol. 2 of a Biography*, (London: Jonathan Cape, 2002).
[2] Flourens, Pierre-Jean-Marie, quoted in Browne, Janet, *Charles Darwin: Voyaging. Vol. 1 of a Biography*, (London: Jonathan Cape, 1995).

more narrowly circumscribed theories of genetics and population dynamics and in the larger framework of evolutionary-historical reconstruction as well.

In this article, I want to examine the concepts of variation and selection in the context of evolutionary theory-as-history. The literature on Darwin has put a certain emphasis on his invention of the concept of "natural selection". Yet Eighteenth-Century "evolutionary" theorists such as Buffon and Charles Bonnet already had developed notions of natural selection which were compatible with mechanicism. There were important novel aspects in Darwin's views on selection in comparison to these previous views on the influence of the environment on organisms, but Darwin's real novelty was in his treatment of variation. It was the product of a series of conceptual transformations which had given rise to the new scientific style of historical reasoning. The work of Immanuel Kant figures importantly in the chain of these transformations.

Historically, the meaning of "variation" was narrowed down when, some decades after Darwin's death, genetics began to provide an explanation of evolutionary phenomena. Genetic reductionism restricted "variation" to variation within the genetic make-up of a population, and saw the source of developing variation in mutations within genes. Variation became thus produced by chance events on the level of the genes.

Only in recent decades, we begin to see a return to Darwin's original, richer conception of variation. On this view, variation also occurs on the level of the organism. In particular, it involves agency on the part of the organism, its capacity in assuming an active role in redefining and negotiating its relationship with its environment, and transforming the environment in the process. Organisms change for their own benefit the environmental conditions under which they are to survive. Habitat choice, learning, and educating offspring in newly learned techniques and habits are among the phenomena studied by the so-called "niche construction" biologists.[3]

When Darwin's evolutionary theory is placed in the larger framework of the new historical understanding of the 19th century, it will become apparent that Darwinian space and time were not Newtonian space and time, the neutral system of coordinates on which we measure the unfolding on laws which are timeless by themselves. Instead, we will encounter the notion "event", giving us a relativistic concept of time, whereas the theater on which the evolutionary play is acted out is likewise subject to change itself. Hence, space is a relativistic notion, too.

Selection

Behind Darwin's concept of natural selection was, of course, a metaphor. Just like plant breeders hand pick individual plants with bigger flowers, or flowers

[3] Odling-Smee, F. John, and Laland, Kevin N., and Feldman, Marcus W., *Niche Construction*, (Princeton: Princeton University Press, 2003).

with interesting abnormal colours, so Nature selects, "rejecting all that is bad, preserving and adding up all that is good", as Darwin wrote.[4] Yet in comparison to a human breeder, nature is blind, as she "cares nothing for appearances." Darwin pointed out that nature "acted" in such a way that in many cases extinction was the result, hence he often used terms like "battle for life" and "struggle for existence".

The necessity of extinction was produced by the "incessant struggle of species to increase in numbers." As is generally known, Darwin was brought to this idea by reading the work of a Cambridge professor of history and political economy, Thomas Malthus. Poverty, war and epidemics of all kinds were the results of a simple law of nature, Malthus had argued, namely that populations tend to increase geometrically while subsistence increases only in an arithmetical ration. Darwin read Malthus in 1838, but before he did, he had already intuited as much while voyaging around the world on board the Beagle. When he started out on his sea travels, Darwin had his mind full of nature's harmony. When he returned, he impressed by nature's harshness. Charles Lyell, the geologist much admired by Darwin, had read Malthus before Darwin did, and he had commented on the ubiquity of violence in nature. However, after completing his voyage around the world, Darwin felt that even Lyell had emphasized too strongly the degree to which organisms were adapted to their environment.

Yet, the basic idea of the environment as an uncompromising sieve, selecting away whatever is unfit for life, was older than the 19th century. Selection was compatible with mechanicism, amounting to some extent, to determinism. Darwin's "natural selection" was heir to an 18th century mechanist understanding of nature's history. Buffon had held that similar environmental conditions would act similarly on living beings, regardless of local and specific historical conditions. Buffon's views were in accordance with contemporary thinking, notably Montesquieu's, on the determining effect of climate on civilizations. The most advanced civilizations, this argument ran, were to be found on moderate latitudes. Here, climates were such that they necessitated humans to work hard and not to develop laziness (as in tropical and subtropical climates). Overly harsh conditions, however, acted as impediments to progress; hence, the superiority of a limited number of civilizations, notably the Western-European and the Chinese.

In a sense, the law-like idea of the 18th century naturalists was that every set of climatic or environmental conditions corresponded with a well-defined collection of organisms. The discovery of fossils in the 17th century had made it clear that the Earth had a history. Robert Hooke and Niels Stensen had found shells and shark teeth high up in the mountains in Italy, and they had noted that the shells were like, but not exactly similar to,

[4] Darwin, Charles, *The Origin of Species*, 1st ed. (1859), J.W. Burrow editor, (New York: Penguin, 1968), especially Chapter 4, p. 135.

existing species. Apparently, these organisms belonged to an era during which environmental conditions were different from ours.

Was there, then, evolution? From around 1700, there was a theory which was often called "evolutionary theory", and it was accepted by almost every naturalist and philosopher. The theory is now usually designated as preformationism.[5] A late version of this theory, advanced in 1769 by the Swiss naturalist Charles Bonnet, explained the occurrence of new species during various phases of the Earth, history through a mechanism which we can identify as "selection". An examination of preformationism will show how far removed it was from Darwin's own evolutionary theory, even in the interesting version that Leibniz had developed.

Eve's Ovary

Bernard de Fontenelle, the long time Permanent Secretary of the *Académie Royale des Sciences*, advanced the following argument in his *Lettres galantes*: "You say that animals are machines just like watches are? Put a dog-machine and a bitch-machine together, and a third small machine could be the result of that; put two watches together during their whole lifetime, and they will never make a third watch".[6] Very similar arguments were made by lots of other commentators as well. The problem was to reconcile life's procreation with mechanicism. The solution was preformationism, which held that all living beings, including those living now, were created by God during the Creation. Once they were in place, the laws of mechanicism could apply to them, but mechanicism was no longer in need of explaining how they had originated. This position had not been Descartes. Descartes himself and several other scientists of his generation had held to epigenesis, stating that an egg started out as undifferentiated matter, and was gradually shaped into an organism by the workings of mechanical laws. Yet no mechanist had been able to discover what laws these were, and epigenesis lost the theoretical attractiveness it once had. Preformationism was at the same time a critique of mechanicism and a supplement to it.

Preformationism held that the egg in a woman (or in any animal of the female sex) contained a fully formed individual. These individuals, insofar as they were of the female sex, would be in the possession of ovaries in which eggs with again fully formed individuals, and so ad infinitum. Reasoning backwards in time, we were all present in Eve's ovaries, created together with Eve herself. (There was, in fact, also a somewhat less popular male variant, putting us in Adam's testicles).

---

[5] By 1790, Kant was skeptical with regard to preformationism, which he indeed designated as '*Evolutionstheorie*'. *cf.* Kant, Immanuel, *Kritik der Urteilskraft und Schriften zur Naturphilosophie*, (Darmstadt: Wissenschaftliche Buchgesellschaft, 1983), p. 376.

[6] Fontenelle quoted in Roger, Jacques, *Les sciences de la vie dans la pensée française du XVIIIe siècle*, (Paris: Armand Colin, 1971), p. 346.

It may seem extraordinary that preformationism had such a wide following, but apart from theoretical considerations, compelling in their own way, there was seemingly solid, albeit somewhat indirect observational evidence. Leeuwenhoek's microscopes had opened up the world of very small living creatures, making visible that a single drop of ditch water held a large menagerie, and his demonstration of the animal-like behavior of male semen had caused widespread admiration and wonder. Swammerdam had shown that the inside of a caterpillar contained a complete butterfly, folded with great ingenuity, just waiting to liberate itself from the caterpillar's skin, and there were similar observations with respect to plants, already formed in the seeds that contained them.

A Dynamic Preformationism

Was it necessary that every germ contained in the womb of its parent one that was exactly similar to it? A century after Leeuwenhoek and Swammerdam, Bonnet developed a theory of the germ that allowed for more plasticity in the way it would express itself as an individual. Bonnet was an ovist, but he thus was able to explain a formative influence by the father. Another set of considerations allowed him to address the transformation of species, which he designated with the term "a sort of natural evolution."[7] First, the germs implanted in successive generations of individuals needed not be exactly similar to each other. Second, changing environmental conditions could select those germs a little different from others. Bonnet was thereby able to imagine that, were we to see the horse, the chicken and the snake at the time of the Creation, as they had just left the creating hands of God, we would not be able to recognize them.

In developing these views in his *Palingénésie philosophique* (1769), Bonnet was heavily influenced by Leibniz, who had commented on preformationism in his *Système nouveau de la nature* (1695), and more extensively in his *Nouveaux essais sur l'entendement humain*,[8] published posthumously in 1765.

Leibniz

In the *Nouveaux essais*, Leibniz had presented an almost fully developed evolutionary theory on the origin of human languages, arguing on the basis of etymological and other examples that all languages derived from an original, primitive language.[9] Why, then, did he endorse the cruder theory of preformationism? Leibniz had been much impressed by Leeuwenhoek's discoveries of very small creatures - they may have led him directly to his

[7] Bonnet quoted in Roger, *Sciences de la vie, op cit,* p. 723.
[8] Leibniz, Wilhelm Gottfried, *Nouveaux essais sur l'entendement humain*, (Paris: Garnier-Flammarion, 1966), Book III, chapter 6, section 23.
[9] *ibid,* Book III, chapter 2, section 1.

famous statement in the *Monadologie* (paragraph 67): "Every bit of matter may be conceived as a garden full of plants or a pond full of fish. But each branch of the plant, each drop of its bodily fluids, is also such a garden or such a pond." [10] With this statement, Leibniz aimed at a more general understanding of (living) matter than the problem that preformationism tried to solve, but it shared with preformationism the assumption that small creatures, below the threshold of human observation, can and do exist.

Leibniz needed preformationism as support for his conviction that living things were indestructible. After all, according to Leibniz, an organism did not really disappear at death, but continued its existence in what remained of it, including in its offspring. Metaphysically speaking, Leibniz wrote, there was no death, and neither was there a totally new created living individual.[11] In contrast, epigenesis in its 17th century version had implied that life started wholly anew with each birth. Leibniz used preformationism, in itself an utterly static theory, to arrive at a dynamic and continuous conception of nature. Yet neither Leibniz nor Bonnet, nor any other 18th century naturalist, had given an indication how dynamics in nature was produced. This account, eventually, Darwin would make with the concept of variation, building on innovations introduced by Cuvier and Kant.

Conditions of Existence and Agency

Georges Cuvier, the anatomist, is not known as an evolutionist. Instead, he is regarded as a "fixist", holding that organisms keep themselves fixed, i.e. maintain their stability *vis-à-vis* their environment. Organisms, according to Cuvier, are able to do so thanks to their *conditions d'existence*, requiring that they be a functional unity adapted to their environment. In his anatomical work, Cuvier had recognized that organisms are internally organized to a plan, and classes of organisms such as the *Vertebrae*, in a word: the higher taxons, would each have their characteristic plan. Changes in this functional unity would be detrimental to the organism, hence the impossibility of evolution. In a famous argument, however, Michel Foucault has developed the view that Cuvier's conditions of existence are the conditions which permit life to have a history. [12] The living organisms themselves are only the transitional manifestations of the ongoing experience of life, functioning as a "wild ontology".[13] We only need slight deviations from ideal adaptation (such as Darwin would provide with the concept of variation), and we have a fully

---

[10] Leibniz, Wilhelm Gottfried, *La monadologie* (Paris: Delagraave, 1881) §67.

[11] Leibniz, Wilhelm Gottfried, *Leibniz's 'New System' and Associated Contemporary Tests,* translated by Woolhouse and Francks, (Oxford: Clarendon Press, 1997), §7.

[12] Foucault, Michel, *Les mots et les choses: une archéologie des sciences humaines,* (Paris: Gallimard, 1966), p. 288.

[13] *ibid*, p. 289.

developed history. The conversion of mechanicism to historicity Foucault found more clearly expressed in Cuvier than in Kant.[14]

Foucault's negative comment on Kant, however, can be qualified. The obvious source for ideas in Kant's work similar to Cuvier's, would be the former's *Kritik der Urteilskraft* (1790). In this work, Kant, too, had stressed that living beings are organized wholes. These cannot be explained by the principles of mechanicism, although without mechanist principles, "there can be no natural science at all", as Kant continued to assert.[15] Instead, we have to assume a goal-oriented character of living beings, both in their internal organization and in the relations between them, such as in the dependency of organisms on other organisms as food. Kant argued that the teleological nature of nature should be understood as an "as-if": we can understand its workings but not how it had come about. Hence while pointing out correctly the shortcomings of mechanicism (although he remained convinced that ultimately a reconciliation with mechanicism would be necessary), Kant denied himself historical explanation.[16]

On this account, the difference between Kant and Cuvier is a difference of degree only. Cuvier also just fell short of acknowledging that nature had a history, even if he had the relationship of organisms with their environment (through the concept of *conditions d'existence*) much more strongly developed than Kant. On this point we might see Cuvier as more advanced than Kant along a similar line of reasoning, and this view can be sustained by the fact that an intellectual lineage has been reconstructed between Kant and Cuvier, going via the "teleo-mechanical school" of Johann Friedrich Blumenbach and his pupil Carl Friedrich von Kieselmeyer. Parenthetically, the Blumenbach school is credited with having revived epigenesis.[17]

Alistair Crombie has drawn attention to a different text, one of Kant's "pre-critical" writings: "*Von den verschiedenen Racen der Menschen*", an address delivered by Kant in 1775.[18] In this work, Kant advanced a number of positive claims and speculations on the formation of the human "races" which, according to Kant, have all developed out of a common stock. He dismissed the possibility of multiple creations, and explained the various differences between the races and varieties of humans as the result of

---

[14] *ibid*, pp. 314-318.

[15] For instance, in Kant, *Kritik der Urteilskraft*, *op cit*, p. 368.

[16] Although Kant played with the possibility of a positive history of nature, the province of the "archeologist of nature" in a manner not unlike he had done previously in his pre-critical writings (*Kritik der Urteilskraft*, *op cit*, pp. 369-371). See also Lepenies, Wolf, *Das Ende der Naturgeschichte*, (München: Hanser, 1976), pp. 37-38.

[17] See Theunissen, B., and Visser, R.P.W., *De wetten van het leven*, (Baarn: Ambo, 1996).

[18] Kant, Immanuel, "*Von den verschiedenen Racen der Menschen*" (1775), www.ikp.uni-bonn.de/kant/aa02/ accessed on 27 March 2008. See also Crombie, Alistair, *Styles of Scientific Thinking in the European Tradition*, (London: Duckworth, 1994), Vol. III, p. 1707.

modifications due to the exigencies of environmental conditions: climate and soil. Their "hidden inner predispositions" enabled humans to develop themselves in different directions, even if this did not (and could not) lead to the formation of new species. People living in the "ice-zone" are smaller because otherwise they could not keep themselves warm. The swollen noses and lips of people living in hot and humid climates would also be responses to environmental conditions. Kant expressed awareness that these examples (and he gave many others as well), were in their specificity just speculations, the scientific merits of which were not for him to decide. However, he insisted on the more general point that once the racial differences had been produced as the result of adaptations to the climate, they would persist, even after secondary migrations to yet different climates occurred. Here Kant offered another speculation: Africa and Asia had been separated in old times by a sea; hence, the Negro and Hindu races were developed so differently. At this point in his argument, Kant expressed the hope that a "history of nature" would once become an independent science, based on understandings rather than opinions (as apparently including his own). The *Naturgeschichte* would not be one of humans only. At several points Kant applied his reasonings to plants and animals, too.

Crombie reads Kant's argument as an ontological precursor to the epistemology of his later "critical" period, in particular his *Critique of Pure Reason*. The similarity would be in the apparently active role humans assume in relation to their environment. "*Racen der Menschen*" deals with the bodily response to whatever their environment or historical conditions offers them. But as Foucault pointed out, Kant's view fits in an 18th century taxonomical view of making visible an "already there"; that is to say, the hidden predisposition.[19]

Kant's epistemological argument of the *Critique*, in contrast, can be seen as genuinely new, and it is to this text that we should turn as enabling us to see the response to the environment as leading to novelty. The *Critique* contained the refutation of the dominant 18th century view as expressed most forcefully by John Locke. How do humans acquire knowledge? Through the senses. The mind is like a wax tablet receiving the imprint of the outer world. Locke also invoked the image of the camera obscura, again with passivity as its consequence. The observer is an involuntary witness to what the world mechanically reproduces inside him.[20] With Leibniz also, the knowledge about the world that monads carried inside them was a passive representation of the world. He went further than Locke, though, by differentiating between monads: their "inner screens" greatly varied among each other.

[19] See Foucault's discussion of Benoît de Maillet in *Les mots et les choses*, *op cit*, p. 166.
[20] See Crary, Jonathan, *Techniques of the Observer: On Vision and Modernity in the Nineteenth Century*, (Cambridge, MA.: MIT Press, 1990), p. 41.

Kant operated a decisive break with representation as a passive mirroring of the world. He placed the subject in the center of how knowledge is produced in his *Critique of Pure Reason*. Through the senses every individual receives a bombardment of impressions, but by themselves they remain meaningless. One mental faculty though, the imagination, is able to filter them, and place them in a meaningful context. The imagination had been known to various eighteenth century philosophers, but they had by and large distrusted it for its flights of fancy. Yet Kant placed all mental faculties under the command of the "I", which he elsewhere also called the "will". As a consequence, he was able to liberate the imagination.[21]

In the end, Kant did not "re-naturalize" the epistemological insight of the *Critique*. On the basis of the *Critique*, he could have advanced the claim that organisms do not undergo the workings of the environment passively. Yet, he did not, and further, he did not come back to his project of a *Naturgeschichte*, outlined in "*Racen der Menschen*" either.[22] Yet the active acquisition of knowledge about one's environment is a prerequisite to responding adequately to it, and this again is a prerequisite to historicization.

Kant did not differentiate between humans or organisms in his epistemological argument in the *Critique of Pure Reason*. All humans intersubjectively share the same mental framework. However, we need to add only Leibniz' view that no inner screen of any monad is alike, to arrive at the differentiated conception of how individuals perceive and act that Darwin needed.

Darwin's Variation

There are several layers in Darwin's concept of variation, each with its own intellectual sources. First of all, there was the plain, observable fact that individuals of the same species are different from each other, in seemingly unimportant but sometimes also in important ways. Over the years, Darwin had collected a considerable number of striking examples which, as he wrote in the *Origin*, would surprise even the experienced naturalist.[23]

The *consequences* of variation were the driving force of evolution. In order for evolution to occur, it was enough that small differences could give the individuals that possessed them a net marginal advantage over individuals of the same species. Darwin combined here a statistical understanding of a population, with individuals possessing fluctuating traits, with economic theory according to which a slight advantage over competitors

---

[21] Engell, James, *The Creative Imagination: Enlightenment to Romanticism*, (Cambridge, MA.: Harvard University Press, 1981); Daston, Lorraine & Galison, Peter, *Objectivity*, (New York: Zone Books, 2007). Kant also mentions the "will" in his posthumously published "*Menschenkunde oder philosophische Anthropologie*" in Kant, Immanuel, *Menschenkunde: nach handschriftlichen Vorlesungen*, hrsg. Von Fr. Ch. Starke, (Hildesheim: G. Olms, 1976).

[22] With the exception of a speculative footnote in Kant's *Kritik der Urteilskraft*, *op cit.*, p. 371.

[23] Darwin, *Origin*, *op cit.*, chapter II, p. 102.

could be just enough to survive, while others just did not. Over very many generations, small differences could become large enough to produce new species.

In the context of historical reasoning, the consequences of evolution express themselves as "events". After an "event", the world is different. The notion of "event" marks the insight that, unlike timeless laws of nature, historical processes go in one direction. "Events" cannot be undone. Under a certain threshold, variations may be small enough to not constitute an "event", but if for some reason variations add up, an "event" is produced, and the course of history goes in one direction rather than in another, making the world as we know it radically contingent. Sometimes, at some places, "events" follow up upon each other in rapid succession; whereas, at other times or at other places few "events" are produced. Time becomes the relational sequence of "events", rather than the neutral background on which timeless laws of nature unfold.

There is an interesting similarity and difference between Darwin and Leibniz on the importance of "events". Leibniz had tried to arrive at a thorough understanding of contingency. He knew that in our present world, Caesar could not have not crossed the Rubicon. In order to reconcile contingency with harmony, Leibniz had developed the notion of "compossibility". Darwin no longer had a need for compossibility. In his view of nature, events may be a source of the continuous production of disequilibrium between species and their environmental conditions. Space becomes the theater on which organisms act out the advantages which variations in their biological constitution had given them - a wild ontology indeed.[24]

Yet what were the *causes* of variation? Darwin's best known answer is that they were the result of automatic statistical necessity. Variations occur at random, by chance. This argument was directed primarily against the idea that the order of nature had a "reason", and equally against the idea that natural laws imply a purpose, as if their workings were under control by a Power.

Nonetheless, in another sense, variation is not totally random. Individuals may change their behavior, which is itself a form of variation, and by doing so, they also change the conditions under which the environment is selectively acting upon them. Here we see the importance of the extended, naturalistic version of Kant's conception of individuals who by force of their varied perceptions of their environment act differently.

Darwin discussed behavior in a chapter called "Instinct", but he pointed out that so-called instinctive behavior is not fixed once and for all. "A little dose...of judgment or reason often comes into play, even in animals

[24] See Deleuze, Gilles, *Le pli*, (Paris: Minuit, 1988).

very low in the scale of nature," Darwin wrote.[25] Some kinds of behavior are clearly acquired, according to Darwin, such as in the case of the magpie, fearful of man in England, where it is hunted, and not at all shy in Norway, where it is not.[26] Darwin acknowledged that instinctive behavior of animals is largely inherited, but, he adds, here we find variation, too.

Darwin had amassed lots of evidence pertaining to migration by animals, exposing themselves to new climates and subsequently achieving (or undergoing) acclimatization. Over the course of several years, Darwin did experiments on plant seeds, testing their ability to survive in extreme conditions, like in sea water. It appeared that very many could, so that extraordinary abilities for migration in the case of plants were also demonstrated.[27] The vocabulary which Darwin uses in discussing migration testify to the active role he ascribed to plants and animals: minute shells "cling" themselves to duck feet, thereby showing "agency". Some species "acquire" a wide range; others have the "capacity to cross barriers".[28] Some fish "reject" certain kinds of seeds after having swallowed them, and thereby transport the seed to other places.

Coda

Political history was the first branch of the new history which originated in the nineteenth century, in the work of Michelet and others, in which no traces were left of the older "mechanist" way of conceiving history, as practiced by eighteenth-century historians such as Edward Gibbon and Johann Winckelmann. Gibbon and Winckelmann saw history unfolding according to an organic scheme: civilizations and styles in art would develop from youth through maturity to a final decadence. For the new historical style of the nineteenth century, history had an open ending, with no pre-fixed goal. The presuppositions needed to write history in the new historical style: contingency, agency, the relationship between an individual and a population, the relativity of space and time, were the very same that Darwin needed to arrive at his "evolutionary theory". Evolution, like history, is made by the very many individuals taking part in it.

To conclude, the Darwinia evolutionary theory also influenced the writing of history as a dynamic intervention in "events" by individuals, not only just natural selection. As shown, the concepts of variation and selection examined within the context of evolutionary theory-as-history, squarely positions living organisms as transitional manifestations of the ongoing experience of life. This broader (and perhaps closer) interpretation of Darwin's concept of variation incorporates "agency" on the part of the

---

[25] Darwin, *Origin*, *op cit.*, chapter VII, p. 234.
[26] *ibid*, p. 240.
[27] *ibid*, chapter XI, p. 354.
[28] *ibid*, chapter XII, pp. 376-377, 391.

organism, as a relationship of responsive negotiation with its environment, effecting the environment in turn. Thus placed in the larger framework of the new historical understanding of the 19th century, Darwinian space of “selection”, and time of evolutionary “events” were not a neutral and static system of timeless natural law. Rather, the notion of “event” with Darwin gives us not only a “relative” conception of space and time, but also a dynamic, interactive one. Time became, with Darwin, the relational sequence of events, rather than the neutral background on which timeless laws of nature would unfold. Space no longer was a neutral void; rather, the organism and the environment were mutually affecting each other, in a system where organisms also had a “will”, acting out the advantages in biological adaption.

The Time of History/The History of Time

Leslie Kavanaugh

Reinhart Koselleck begins the preface in his volume of essays, *Futures Past* with the question: "What is historical time?" In thinking an historical time that also takes into consideration that time epochally extends itself into territorial space, the following historians and philosophers will be examined: Bloch, Koselleck, Foucault, and Poulantzas. One of the primary intuitions of these thinkers is that there are hybrid spatialities and multiple temporalities, each one heterogeneous and yet interrelated with the others. Specifically, the question to be asked is: how could the notions of "non-contemporaneous contemporaneity" and of "mixed disparate temporalities", begin to problematize history as a spatio-temporal field, comprised of "layers of time", and dynamically changing? Thus, the task of this essay will be to explicate some of the spatio-temporal layers that could conceivably be of import to the considerations of the unfolding dynamic of the phenomena of space-time.

"All history is made in the present"[1]
"Time is the order of existence of those things which are not simultaneous"[2]

Problematically, space and time have historically been considered as separate categories. Obviously, when examining various historical instances, this approach yields up rather reductionist models where territories as spatiality – whether "post-colonial" or "capitalist/democratic" – need be privileged over a linear conception of historical time. This formalist model can lead quite simply to misconceptions as to the complex nature of the unfolding social, political, and economic landscape. Increasingly, theoretical models have been developing in order to come to grips with this complexity. These models are variously termed, "networks", "meshworks", "non-contemporaneity", space-time problematics, "globalized spatial/temporal fix", "spatial and temporal matrices", etc., in an attempt to understand the operations of cultures, whether western or non-western, beyond an isolationist or rigid paradigm. Specifically, the problem concerning the modernist or capitalist development in a global environment will be examined using these new

[1] The "present" in this case being 23 November 2005 at 3:19 p.m., spoken at the occasion of the Alexander Humboldt Lectures in Human Geography, Radboud University, Nijmegen, The Netherlands. I am grateful to Professor Harootunian for introducing me to the work of Nicos Poulantzas upon this occasion and providing me with an advance copy of the following publication: Harootunian, Harry, "Remembering the Historical Present", *Critical Inquiry*, volume 33, number 3, Spring 2007.
[2] Leibniz, Gottfried Wilhelm, "The Metaphysical Foundations of Mathematics (after 1714)" in *Leibniz: Philosophical Papers and Letters,* edited and introduced by Loemaker, Leroy E., (Dordrecht: D. Reidel, 1956), p. 666.

models that attempt to avoid a hierarchical superstructure and a separate layer of historical determinism. Rather, in order to come to terms with the complexities of shifting territories and multiple temporalities, the current emerging models will be elucidated as to their usefulness in understanding space-time as applied to cultural/social/economic concerns.

The Layering of History as a Matrix: Harootunian

In Harry Harootunian's reading of an historical framework, specifically in the work of Lefebvre, Kosellek and Poulantzas, he emphasizes that they "have all insisted on…the importance of trying to show how time, usually [in the form of] history, have interacted with a specific space to form an identifiable relationship."[3] Although Harootunian is primarily an historian, specializing in Asian Studies/Japan, his later work begins to provide a theory for historical time that attends to what he calls the "space-time problem". Historicity, in general, has paradoxically always given preference to space over time, but Harootunian takes up the problematic in light of the increasingly urgent calls in the 20th century for a re-thinking of notions of time – including in this case epochal or historical time – in relation to or with concepts of space. As a scholar specialized in Japan specifically and post-colonial history generally, a problematizing of a new space-time dynamic would obviously be of great import. Most interestingly, the reading of history that Harootunian proposes might provide a viable alternative to linear concepts of the unfolding of historical time.

According to Harootunian, with the project of modernization of societies, time is determined as "progress" (usually with outside and often unwelcome intervention) until such time as societies "converge" into a single market and all differences are evened out. This modernization project however, generally means that the hegemonic societies determine the terms of this progress, and the speed at which it occurs, which *de facto* comes down to the Westernization of global culture. All "other" cultures/economies were "mapped onto [this] evolutionary trajectory".[4] Indeed, a condition of equal terms between "capitalization" and "democracy" insured the voracious spread of capital markets for primarily, although not necessarily, Western capital. In fact, "democracy" became synonymous with "capitalism". The purpose was in fact the creation of markets, and the role of culture took on a mere custodian function. The "progress" defined by modernism had a linear

---

[3] Harootunian, Harry, "Some Thoughts on Comparability and the Space-Time Problem", *Boundary* 2/32:2 (2005): 24. Harootunian takes up and treats thinkers such as Marx, Lefebvre, Bakhtin, Poulantzas, Bloch, Kosellek, Lukacz, Halbwachs, Tetsuro, Weber, Fabian, Postone, Jameson, Simmel, Althusser, Harvey, as well as Foucault, de Certeau, Arendt, and Benjamin in this essay.

[4] Harootunian, "Some Thoughts", *op cit,* p. 30.

temporality – proceeding from underdeveloped capital markets (third-world or foreign economies) to fully "integrated" and "evolved" growth economies. Yet one must remember that this notion of time as historical progress was not only linear, but also uni-directional, a time measured only in terms of its spatial distance to the present. The intention was always that the "under-developed" economies accept the dissemination of capitalism, democracy, and what was in effect Western culture until all differentiation was converged into a seamless flow of capital. It was never the intention, conversely, that the capital accumulation would remain in these countries, or that these "other" cultures would influence the hegemonic cultures in either accumulation or labor flows.

After the Cold War, according to Harootunian, the project of social engineering or "progress" in a modernist sense became unnecessary, emptied out, pointless. A sort of free-fall condition occurred where time looped back on itself, unable to find its direction. No longer did the old binary oppositions exist: between capitalism and communism, between nature and culture, between rational economic plan economies and liberalism – either to uphold them or to critique them. Without opposition, historicity could no longer "progress"; that is to say, one then had to ask: "progress towards what?"

Yet the withdrawal of confidence in the binary categories of capitalism/communism, or West/Non-West developing countries, now merely bifurcated into a global blanketing of a "war on terrorism". This appeal to war then became the undifferentiated narrative of evening-out or reductionism the likes of which modernism could only dream of. In fact, the "war on terror" became the new unifying category in the absence of any other alternative, including oppositional alternatives. Finally, seemingly, all obstacles were removed to global capital.

As a result, a deep suspicion of "utopia" or indeed futuricity itself began to be heard. This suspicion not only sought to deconstruct the linearity of time as modernist "progress", but also indicates a kind of lack of nerve. The "market" seems to be the only remaining transcendental principle to be universally applied. Not only were the Utopian projects such as Marxism rendered (at least in its Leninist/Stalinist/Maoist forms) completely violate, but the critique generated from Marxist philosophy, and so-called "critical theory" was swept from any kind of serious further consideration as a possible way forward out of the impasse. In addition, the suspicion of a utopian project, in the form of a critique of 17th-century modernism/Enlightenment ideals; [5] indeed, forms of rationality and the oppositional structure of epistemology and ocular-centrism, was so

[5] *cf.* Dupré, Louis, *Enlightenment and the Intellectual Foundations of Modernism* (New Haven: Yale UP, 2005).

completely taken up in the post-War II period by among others Foucault, Lyotard, and Harvey that any "grand narrative" of history seemed untenable. Although Harootunian contends that capitalism only entirely produces unevenness and difference,[6] I would contend that in the wake of the so-called end of the Cold War, the complete self-proclaimed triumphalism of the Americanized version of capitalism/democracy (and precisely the conflation of these terms) has conspired to remove difference into the same, diversity into hegemony, and multiplicity into the mere many/multiple. All becomes equalized, and flattened.

This concept of "unevenness" comes from Marx's material conception of history where civil society is fundamentally determined by its political economy. In a process of the development of capitalism, a transformation takes place from the former feudal economies into the exploitation of labor and means of production by an increasing few. The capitalist system per definition is an alienation of labor with the means of production. The laborer is free only to sell his labor on the open market. As a direct result of the development of capitalism, according to Marx, "the history of this expropriation, in different countries, assumes different aspects, and runs through its various phases in different orders of succession, and at different periods."[7] Thus, capitalism necessarily produces unevenness. Yet paradoxically, capitalism is a voracious beast; it must grow. Whilst capital accumulates in an ever tighter concentration, new markets must continually be found in order to create surplus value. This increase is accomplished by either increased productivity, lower costs, or broadening markets. As a consequence, labor costs must be forever reduced, material resources exploited more efficiently, and consumers must be persuaded to consume beyond what is strictly necessary. All combined, all peoples get "entangled in the net of the world market...Centralization of the means of production and socialization of labor at last reach a point where they become incompatible with their capitalist integument."[8]

Although in different spheres, Marx explains, the rate of growth progresses at differing rates in an uneven pattern, capitalism must necessarily subsume all to its logic. Capital flows in ever-insistent search for expedient labor, cheap resources, and exploitable markets. Consequently, even though capitalism produces unevenness, in fact capitalism eventually levels-out and equalizes/totalizes, encompassing dissimilar, heterologous relations in its

---

[6] Harootunian, *op cit*, p. 33.

[7] Marx, Karl, *Karl Marx: Selected Writings*, edited by David McLellan (Oxford: Oxford UP, 2001), p.523. See also in the same volume, the "Preface to A Critique of Political Economy", pp. 424-7.

[8] Marx, *ibid*, p. 525.

political economy. Capitalism becomes a superstructure necessarily encompassing all in its path.

The largest temporal abyss, the most extensive contemporaneous contradiction is indeed the specter of the proletariat that indeed no longer exists except mythologically, as well as the disengagement of capital from property relations and the "means of production". Labor can no longer be organized into a class called the exploited "proletariat" – even though persons are exploited more than ever, in increasingly inhumane and dangerous ways – rather, labor is nomadic, fluid, pooling into schools like fish, moving from place to place where needed. Similarly, capital itself is invested – not in means of production or facilities or human capital as a long-term investment – but as a kind of strike-force mentality. Far more cost-effective is it to abandon production sites and to resituate the locus of labor and means of production than a long-term capital commitment to a specific place, or product, or material resource base. This nomadic, hyper-capitalism has obvious and devastating effects for the stability of communities, and even countries. Indeed, the "Now" that both Harootunian and Ernst Bloch speak of as non-contemporaneous, is not estranged merely from the past, but from itself: the "Now" is already tomorrow.

Yet, as Harootunian says, "All history is made in the 'present'." And it is Harootunian's conception of history then – history as a space/time concerned with not only the past, produced in the present – but a history that has contained within it its own futuricity, that can provide a way of thinking forward out of the trepidation of making a wrong step, of again setting up contentious contradictions and oppositions - time that is decidedly not a conflation, but a configuration; time not though of as a simple chronological scheme of past, present, and future. Yet time has seemingly in the advent of the effective collapse of communism, has not only radically thrown temporality into "an endless present", but erased in advance any conception of futuricity. For where will we go if we no longer believe in "progress"?

In the end, can we have a "history" as a universal or transcendental, in fact, a-historical category? We need a new concept of historical time that accounts for the unevenness of global capital in the wake of the breakdown and reconfiguration of empire, and replaces the universal/transcendental historical with a more "uneven" albeit more reflective of the historical present, a folding-in of mixed temporalities, a co-evality of the past, present, and future. Instead of the common practice of examining history by marking off historical events within a given location, we might consider what it would look like to consider history as a multiplicity of hybrid temporalities. As

Reinhard Kosellek emphasizes in *Futures Past*,[9] a categorical confusion often occurs by spatializing time and temporizing specific locations. Instead, we might begin to think of the measure of time as something intrinsic to the thing itself. If each thing has its own time, an "internal time", then perhaps the jump to conceptualizing multiple times coexisting is not so great. Consequently, we would have a temporal model of an infinite flow of individual times coexisting on a plane of immanence, thereby privileging the local/individual whilst acknowledging the connectiveness of the whole space/time structure.

Space and time, then, are not a background upon which, or behind which, events occur and borders are drawn; rather, space and time are a relationship. We have a tendency to forget that we speak of a specific locality only at a moment in time. Space, too, has its own time.

If one makes an historical survey, space and time have also over time had an historicity, a destiny that has been anything but consistent. Differing temporalizations as well as spatializations have determined our historical relatedness. What is now important – in this epoch – is to think the difficult space/time of the present instant and how indeed to formulate our relatedness to some futurity. Or, as Bergson has said, every contemporaneity always includes with it some "*memoire*", or recuperated past that is folded-into the ever-present. Who, then, is afraid of the future?

Ernst Bloch: Heritage of our Times

In responding to the moral and material devastation of the First World War, Ernst Bloch tried to conceive of historical time in different ways in order to come to terms with his contemporary situation. Ernst Bloch diagnosed the mixed temporalities, or the layers of epochal time in the Germany situated historically between the two world wars. The scars of the first war were pre-eminent, while the second was a mere intimation. Specifically in his essay entitled, "Summary Transition: Non-Contemporaneity and Obligation to Its Dialectic" of May 1932, he identified various streams of time, both archaic and contemporary, that co-existed alongside each other in an untimely manner. "Not all people exist in the same Now", he poignantly states.[10] Even though people are obviously living in the same historical period of time, in this case Germany between the two world wars, he analyzes three social classes of the youth, the peasant, and the middle classes in order to ascertain how in fact Hitler could rise to power.

---

[9] Koselleck, Reinhart, *Futures Past: On the Semantics of Historical Time*, translated and introduced by Keith Tribe (New York: Columbia UP, 2004).

[10] Bloch, Ernst, *Heritage of Our Times*, translated by Neville, and Stephen Plaice (Berkeley: University of California Press, 1991), pp. 97-110.

Each class, in its own way, carried an unresolved past along with it into the contemporary situation, setting up an untimely temporality, or what Bloch calls "non-contemporaneous contemporality". The youth, disaffected and without hope of work, are easily manipulated with empty promises of meaning embodied in a fabricated prehistoric mythology. They are seduced into nationalism, whereas the peasantry genuinely is embedded in the conventions and customs of the land, unable in the end to accept their displacement by machine and redistribution of agricultural lands. Peasants, however tied to each other by a seasonal sense of time as well as the appeal to the eternity of the land, failed to organize as an effectual political group. Similarly, the middle classes are paralyzed into passivity precisely because they do not wish to see themselves as a class at all – even as a class opposed to the laboring classes. Mistrustful of the future, they are according to Bloch, "losing time", deflecting their resentment and frustration of the loss of their fortunes onto the prospect of racism and an archaic ideology of false *hubris*, unable to accept the fact that they too have become mere commodities. In the wake of the economic crisis of Germany in the 1920's and 30's, non-integrated remnants of older economic and political systems existed in a heterogeneous, non-incorporated fashion. Ineffectual politically, socially, and economically, all the classes eventually obey. "Believing, obeying, fighting, are these the fascist virtues?"[11] Certainly, the greatest of the three is obeying.

Filled with impotent rage, they are "out of step" with the barren "Now". The only recourse that they have is to resurrect the archaic. Yet this folding-in of the unrecoverable and indeed non-existent past results in the slippage of time into the condition of non-contemporaneous contemporality. Whilst the world has become entirely commoditized and reified, the specter of history haunts the "Now". In an often contradictory and ill-resolved fashion, non-contemporaneous remnants of the past "haunt". They attempt to fill up the vacuum caused by the non-contemporaneous nihilism, the incompatible contradictions inherent in the economic developments of late capitalism, fill it up with romantic myths of national identity, the intoxicating comradery of war, authoritarian state structures, "Christian norms and values", and other rash desperate measures. Yet in spite of these attempts to cover the abyss of nihilism, a rift, a hiatus arises, according to Bloch, "which can [only] be consoled or filled with mist."[12] The non-contemporaneous contradictions are ineffectual precisely due to their "mistiness", their diffusion, their inability to gather into a historically decisive direction; the consequence of which is not only a contradicted present, but a "prevented

[11] Bloch, *ibid*, p. 104.
[12] Bloch, *ibid*, p. 110.

future". The present, the time-entangled "Now", cannot gather itself up in order to project itself positively into the future. The mist encompasses all while history "haunts" the present. Nevertheless, Bloch states:

> The worker who recognizes himself to be a commodity reveals at the same time the equally frenzied and spooky commodity character of capitalist society, without him – as a new class – being fooled with old stuff, without even the "person" or the "life" which he opposes to reification already being historically determined anywhere. The proletarian as the self-dissolution of bourgeois society, indeed of every class society whatsoever, is the subjectively and objectively personified contradiction of contemporaneous society itself, and his revolution – as the fruit of the dialectical recognition of contemporaneous contradictions – complains about no figures and memories, nor at first even any contents of the past at all, but activates purely the future society with which the present one is pregnant, towards which the anarchies and nihilisms of the present one seek to effect their sudden change.[13]

This revolutionary person has "Today", as opposed to the non-contemporaneous "Now". This "Today" neither regrets the past nor is haunted by it. Through recognition of the very character of contemporaneous contradictions, he escapes the mists: alienated labor, uneven material development, "the fetish of the commodity, the instability of nothingness"[14] and recovers the possibilities of a potential future. The foundations of non-contemporaneous contradictions for each class – the youth, the peasant, and the middle-class – had been an unresolved relationship to the past and a belief in a fairytale of the "good-old-days". Yet the unfinished past must be recognized as simply that – irresolvable, haunting, not yet completely expressed; rather remaining multi-layered and interwoven heterogeneous temporalities.

Everyone, Bloch had said, did not exist in the same "Now". Yet maybe all classes can occupy the same non-contemporaneous present, a present pregnant with possible futures without attempting to "master" or resolve the contradictions. Just letting them be. So, in opposition to Hegalianism, no resolution occurs; they remain multi-temporal and multi-spatial and multi-class. No Marxist revolution can solve the problem of multi-layered development, and Bloch acknowledges that the earlier

---

[13] Bloch, *ibid*, p. 111.
[14] *ibid*.

contradictions or uneven developments or "uncorrected elements" can never be completely resolved into late capitalism; rather he argues for practical interventions. Yet, I would suggest, no totality can absorb all contradictions, nor should it attempt to absorb everything into itself; rather, a constellation of contradictions can be constructed, and in this way a certain kind of reconciliation. Yet, do we, in the end, stand on a foundation of mist?

Gurvitch's Social Time

George Gurvitch was one of the most influential scholars of post-war Europe, inheriting the chair of Emile Durkheim at the University of Paris, Sorbonne, and later teaching at the New School, New York. Notably, Gurvitch preceded Foucault in the establishment of knowledge as a social construct, rather than universal, a-historical, or a priori. Having witnessed firsthand in the country of his birth the Russian Revolution, he knew as no other the impact of the group collective upon political events. Although Marx would be a profound influence upon his work,[15] so too would the phenomenological movement sweeping France, much due to his scholarship, in order to explain the relationship between the individual and larger groups, tribes, economic classes, and societies.

Gurvitch briefly treats the subject of space in *The Social Frameworks of Knowledge*.[16] In this work, he is primarily concerned with the establishment of the methodology necessary for the sociology of knowledge. In a section entitled, "Types and Forms of Knowledge", he explicates the knowledge of the external world that is, following Kant, constructed from specific places and times. Undeniably, perceptual knowledge is predicated upon some kind of perception of space and time, the collective being socially constructed by economic classes, kinship groups and societies.[17]

Surprisingly, Gurvitch sees space and time as "obstacles" to group ambitions and aims, yet through struggle groups come to define their social framework. Obviously, perceptions must be collectively held; a singular perception by one individual runs the risk of being termed deviant, or worse, a state of psychosis. Yet Gurvitch, following Bergson, states that although times are varied and individual, space is shared by the collective, as an "outside" of spatial plurality. "The space wherein the external world and the social world are placed, as a reality independent of all consciousness, can be

[15] See for example: Gurvitch, Georges, *Dialectique et Sociologie* (Paris: Éditions Flammarion, 1962); and Gurvitch, Georges, *Le concept de classes sociales de Marx à nos jours* (Paris: Centre de documentation Universitaire, 1954).
[16] Gurvitch, Georges, *The Social Frameworks of Knowledge*, translated by Margret A. Thompson, and K.A. Thompson, (New York: Harper & Row, 1971).
[17] Gurvitch, *ibid*, especially 22-26.

only a single space."[18] Although each individual would have their own perception of space, this space is still one single space constituting one mutual reality. For Gurvitch, time has a different ontological status. Time exists independently from individual perceptions, albeit a multiplicity; time is a shared history. Gurvitch, ever the sociologist of categories, describes space as either a relationship between the individual and the group to the subject, or space as a "function of perceived space".[19] He then applies the categories of autistic, egocentric, projective, and prospective space to the first spatial criteria; and subsequently, the categories of diffuse, concentric, and expanding/contracting space to the various forms and types of knowledge. As such, the diverse variety in space and time, although constituting one coherent picture, express themselves within social frameworks.

The topic of time is much more extensively treated in Gurvitch's influential *The Spectrum of Social Time.*[20] In this relatively short work, Gurvitch carefully explicates the study of time and its relationship and importance to sociology. Passing in review are Jean Piaget's theories on psychology; Henri Bergson's concept of duration; Fernand Braudel's conception of history; M. Lecomte de Nouy's biological time; Marcel Mauss's ages of man, and Paul Fraisse's research on psychology and perception. Most important for our discussion is the fact that Gurvitch considers time as "social" and consequently collective and individual in all its variety. Of course, the complexity of time seen as multiple very quickly becomes a methodological quagmire. Historians have always had to contend with these issues, but Gurvitch, as a sociologist, argues that time as a social manifestation needs other theoretical frameworks. In an attempt to come to terms with the dynamically changing phenomenon of social time, Gurvitch contrasts the methodologies of history with those of his framework:

> Historians describe the particularly rich variations of historical time and their unifications, while sociologists furnish the historians with indispensable conceptual tools to enable them to grasp the concrete expressions of the multiplicity of time and to remind them of the limitations of the continuity between the different time scales to which they are propelled by their reconstructive, projective and singularizing method. Neither history nor sociology possesses a monopoly of the study of social time; they have complementary approaches to it.[21]

---

[18] Gurvitch, *ibid*, p. 24.
[19] Gurvitch, *ibid*, p. 25.
[20] Gurvitch, Georges, *The Spectrum of Social Time*, translated by M. Korenbaum (Dordrecht: Reidel, 1964).
[21] Gurvitch, *ibid*, p. 7.

Although macro-sociology also attempts to explain global events that are cross-tribal, in Gurvitch's social framework of time, he cautions against attempts at unification, explanations of phenomena that have a tendency to reduce the complexity of expressions of social time to a totality. Indeed, Gurvitch celebrates the complexity of the multiplicity of social temporalities. Each social framework has its own rhythm, qualities, speed, migration, and duration. Part of the aim of society as a whole is precisely to unify, or to order, or to attempt to tell a coherent history, about all the manifestations of social time, in all their "multiple manifestations".[22] "A society cannot survive without some unification of the plurality of its social time, yet special problems emerge in the very effort to achieve some consistency in this matter. Every all-inclusive society tries to establish a unifying hierarchy of social time".[23] For Gurvitch, the major distinction between time as an object of study for the historian, as opposed to the sociologist, is that for history time has already passed or elapsed, whereas time for the society is constantly emerging.[24]

Gurvitch, then, defines time as the "convergent and divergent movements which persist in a discontinuous succession and change in a continuity of heterogeneous moments".[25] Now, one may wonder whether this definition elucidates or confuses, but let us take one term at a time and see if we can unravel the "multiple manifestations of time". Emphatically, Gurvitch rejects other historical considerations in the philosophy of physics, including Aristotle, Kant, Hegel, and Einstein. Also inadequate to the theoretical framework of social time are various theories of psychology and perception, including Bergson, Piaget, Bachelard, and Fraisse.

The times of distinct scientific frameworks are also not reducible to one single theory of temporality. Obviously the quantum and macro-cosmological, the biological and evolutionary, the chemical and geological time references are disparate. In social time, by contrast, "all these levels permeating each other and struggling with each other are animated by the varied pulsations of continuity and discontinuity."[26] In the various modes of participation in human life, whether scientific or experiential, the different times rub up against each other, struggling for effectiveness in the structure of social life, creating a frisson between spontaneous and established structures. For Gurvitch, the incongruent yet distinct elements are gathered together in one teaming fluctuation of social interaction. Some elements are

---

[22] *ibid*, p. 13.
[23] *ibid.*
[24] *ibid*, p. 35.
[25] *ibid*, p. 18 (emphasis in the original).
[26] *ibid*, p. 29.

immanent to the system; some elements are vying for hegemony; yet all are part of the eternal flow of human life.

Furthermore, Gurvitch proposes a scheme of heterogeneous times that form a composite social framework. This scheme was also established in the earlier work of *Déterminismes sociaux et liberté humaine* of 1955,[27] but is further elaborated in *The Spectrum of Social Time*:[28]

1. Enduring Time
2. Deceptive Time
3. Erratic Time
4. Cyclical Time
5. Retarded Time
6. Alternating Time
7. Time in Advance of Itself
8. Explosive Time

These diverse times are not unrelated. Although they do not form a totality or a unity, they do comprise a related categorical similarity, forming one interwoven continuity. Enduring time can be described as slow, protracted time, almost imperceptible because it is the time of historical periods. Yet within this continuity, "deceptive time" punctuates the enduring time with sudden events or crises. In this social time frame, situations occur that radically alter the direction and movement of time, but deceptive time is different from the next category of time, "erratic time", in that this time is chaotic and unpredictable in its irregularities. In erratic time, no set pattern or rhythm can be discerned; and therefore, is generally seen as dangerous and disruptive. In contrast, "cyclical time" is epochal and measurable. Cyclical time refers to the standard of time involved in the movement of planetary bodies, and thus an ancient conception of time that is not linear, but cyclical. "Retarded time" is the time that is awaited on the far horizon of events, such as messianic time, or aguishly awaiting peace. "Alternating time" is similar to erratic time, but periods or rhythms, delays and advances can be perceived or abstracted from events.

Most important in modernity is "time in advance of itself", or "time pushing forward", and refers to capitalisms' "uneven development", or a utopian impulse to orient the present towards a possible future. The last category of time that Gurvitch proposes is that of "explosive time", whereby the page of history is turned as it were, such as the Russian revolution, or the fall of the Berlin wall. Explosive time is distinguished from "deceptive time"

---

[27] Gurvitch, Georges, *Déterminismes sociaux et liberté humaine vers l'étude sociologique des cheminements de la liberté* (Paris: Presses U. de France, 1955).

[28] Gurvitch, *The Spectrum of Social Time*, pp. 13-14, and further elaborated on pp. 31-33.

in that these are "acts of collective creation",[29] whereby forces that were perhaps delayed or retarded spontaneously come to fruition and massively change the course of human history. Social time, consequently, is fluid and non-reducible to a hegemonic or measureable time; a time that is a collective of disparate times, enduring, yet not a homogeneous unity.

Gurvitch then uses this categorical scheme of social time in order to study and analyze various historical configurations, such as various social groupings and social classes as well as types of historical societies. Most important for our discussion is the techno-bureaucratic time of modern contemporary society. In the case of modernity, time is directed toward a utopian future; time is "out of time"; time is a force used by various governmental bodies or charismatic leaders in order to push time forward. As such, these bodies or persons form a link "in the chain of immanent and transcendent movements".[30]

In modernity, as Ernst Bloch has also suggested, each social class bifurcates into its own specific time. Time leaning forward as it were, is a time in which "competitive capitalism seems to favor the expansion of production, the acceleration of technical inventions, [and] the promotion of a heightened level of consumption even among the proletariat".[31] Just as with space that must "be produced" in capitalism, time must be speeded up. Time is never present, never "now"; rather, time is the next market to be conquered, the next labor-saving device to increase productivity, the next material to be exploited. As such, competitive capitalism not only becomes "global", but necessarily unifying, totalizing, and subsuming. Yet the social time scheme of Gurvitch also acknowledges that, even though various social classes or historical elements will strive for control over social time, various temporal elements will escape unification – long awaited, individual or collective social times erupt into the flow of time, decentralizing control, and resisting the impulse to the absolute fusion of time. Nevertheless, Gurvitch concludes, "creative time and the time pushing forward can serve evil as well as good....All that we wish to emphasize here is the great multiplicity of social times, their many possible combinations, and the time scales they form in relation to global societies of different times."[32]

Most notably, for Gurvitch's notion of the spectrum of social time, the diverse times are productive and creative, a force to be interwoven into the social fabric, yet retaining a separate distinctness. The various series of time change and pull apart, but also make up a collective of social expression that is a sinuous and fluctuating heterogeneity. As such, the methodology that

---

[29] Gurvitch, *ibid*, p. 33.
[30] *ibid*, p. 110.
[31] *ibid*, p. 137.
[32] *ibid*, p. 151.

he developed in sociology, the "discontinuitic typological method",[33] is more effective because it preserves the plurality, whereas historical methodology, in trying to abstract from events some kind of continuity, must necessarily destroy the immanent succession of human experience. Vitally, for Gurvitch time is not something undergone, but creatively produced. Society "gives birth to time".[34]

Koselleck's Non-contemporaneous Historical Time

Historically, science, including the study of history itself had tried to be universal, and in fact, a-historical. Paradoxically, history as a transcendental worldview became disengaged from the historical unfolding of events. Obviously, history unfolds into the past, yet our reflection upon it, our coming to terms with it, also takes time, so that we always have a backward access to history. So in a sense, history becomes de-historicized; history becomes "timeless"; historical time becomes "u-chronic".

Reinhart Koselleck, as a consequence, attempts to develop a methodology that takes as its first premise that historical time unfolds in a given place in an entirely different way than that of the natural sciences. Chronology, or the attempt to describe history from a transcendental perspective, leads to empty categorization and periodization of historical events. Historicity (*Geschichtliche Grundbegriffe*), in contrast, accounts for the complex and multilayered events that are causes, consequences, and most importantly future causes of historical interventions. With a strict chronology, history not only mendaciously attempts to mimic the abstract time of the natural sciences, but also finds it impossible to "establish causal relationships between the temporal layers"[35] in history.

History does not proceed in a linear fashion, rather it is a complex system of layers: some events pregnant with meaning, others insignificant; some deviant, others a more constant theme. The historian, in fact, must walk a fine line between the God's-eye-view that eradicates difference and causality, and the relativism that always threatens to be the result of a much too detailed inspection of "little histories".[36] In addition, the historian is in the position to be not only engaged with the present; indeed, unavoidably so,

---

[33] *ibid*, p. 152.

[34] *ibid*, p. 30.

[35] Koselleck, Reinhart, *The Practice of Conceptual History: Timing History, Spacing Concepts*, translated by Presner *et. al.*, with a forward by Hayden White (Stanford, CA: Stanford UP, 2002). This work is a collection of essays dealing chiefly with the problem of the discipline of history and its historical antecedents. In the German language, the theoretical ties between the layers of history are more pronounced: layer (*Schicht*), and history (*Geschichte*).

[36] Koselleck cites Thucydides: "Not every testimony … is equally worthy of being reported. He hierarchizes his sources in order to capture history's immanently demonstrable, naturally similar, partly tragic, and always self-concealing potential for conflict." Koselleck, *ibid*, p. 67.

framing his questions and his inquiry in the light of contemporary concerns, yet also "transcending" conceptually contemporaneous time in order to draw out the significance of events.[37] In the end, both "scales" in the unfolding of events are significant, comprising the multi-layered richness of temporal continuity that is necessary to make comparisons, and to speak about the "place" of man in historical time.

Without the time of the natural sciences providing a measuring stick, and without a linear conception of history, another theory[38] must be actualized in order to not only tie together synchronic events, but also provide a substratum that accounts for succession and duration. A difficult task indeed, for historicity must provide an account for both the intersubjectivity of various historical actors, and the interconnection of events. Koselleck describes this difficult paradox of multi-layered time as what he terms, "non-simultaneity of the simultaneous" (*Ungleichzeitigkeit des Gleichzeitigen*).[39] In *Futures Past*, he outlines three main modes of temporal experience:

> 1. The irreversibility of events, before and after, in their various processual contexts.
>
> 2. The repeatability of events, whether in the form of an imputed identity of events, the return of constellations, or a figurative or typological ordering of events.
>
> 3. The contemporaneity of the non-contemporaneous (*Gleichzeitigkeit der Ungleichzeitigen*). Within this temporal refraction is contained a diversity of temporal strata which are of

---

[37] Indeed, Koselleck admits that teleology is the presupposition of most historical practice. "By making a statement about an event and by adding that it had to have happened, I vindicate for this event a necessary causal chain – a necessity that in the end derives from the omnipotence of God, in whose place the historian is acting." Koselleck, *ibid,* p. 13.

[38] Koselleck acknowledges that others, for example Herder and Goethe, in the Enlightenment, had indicated the multifaceted nature of time. He cites Herder's critique of Kant's *Critique of Pure Reason*: "In actuality, every changing thing has the measure of its own time within itself....No two worldly things have the same measure of time....There are, therefore,...at any one time in the universe innumerably many times." Herder, Johann G., *Metakritik zur Kritik der reinen Vernunft* (Berlin: Aufbau-Verlag, 1955), p. 68. Cited in Koselleck, Reinhart, *Futures Past: On the Semantics of Historical Time*, translated and introduced by Keith Tribe, (New York: Columbia University Press, 2004), p. 237, and in a slightly different translation cited in Koselleck, *The Practice of Conceptual History*, p. 111.

[39] Unfortunately there are two variants of this difficult term. The translation of *Ungleichzeitigkeit des Gleichzeitigen* as "non-simultaneity of the simultaneous" is from Todd Samuel Presner in *The Practice of Conceptual History*. The term (*Gleichzeitigkeit der Ungleichzeitigen*) has also been translated by Keith Tribe as "contemporaneity of the non-contemporaneous" in *Futures Past.*

> varying duration, according to the agents or circumstances in question, and which are to be measured against each other.[40]

These modes of temporality have several consequences. Firstly, even though the temporal strata are separate, they do influence each other. Although arriving and departing at various speeds, they are punctuated by various events that tie them together. Also, time itself can become an agent in pushing the layers, in the case of, for example, modernity where the orientation was no longer the past, or even the present, but facing some utopian promise of a progressive future. And lastly, the binding force operating on the temporal plane is the subjective perspective (*Standortbindung*)[41] of human experience. "All historical knowledge is locationally determined and hence relative."[42]

Although obviously each person experiences events from their own singular point-of-view, this fact is not to imply that there are as many temporal strata as there are individuals. History is also temporally constructed in the present, yet at multiple distances from the present, with disparate relations to the past, the present and the future, the various strata or layers of time becoming non-simultaneous in their contemporaneity.

In bringing to the fore the inadequacy of both the chronology of the natural sciences, and the epochal periodization of historical studies, Koselleck introduces a different conception of temporality due to its orientation; namely, the "space of experience" and the "horizon of expectation".[43] On the one hand, instead of the past-present-future triad, Koselleck proposes the categories of (memory)-experience-expectation.[44] Undoubtedly, these are categories derived from human, lived concrete reality. On the other hand, Koselleck argues that "the categories of 'experience' and 'expectation' claim a higher, or perhaps the highest, degree of generality, but they also claim an indispensable application. Here they resemble, as historical categories, those of time and space."[45] Yet, most importantly, none of these categories is meaningful without the specific situatedness which is humankind temporally positioned, embedded in historical time. "Experiences

---

[40] Koselleck, *Futures Past, op cit,* p. 95.

[41] Koselleck, *ibid,* footnote 1, p. 293: "*Standortbindung*" can be translated as "perspective", but also in German indicates a specific "point-of-view", or a determinative temporality grounded in a specific place experienced by the "standing" or position of a human subject.

[42] *ibid,* p. 129.

[43] *ibid,* p. 255 *ff.*

[44] *ibid,* p. 255. Koselleck does not offer a theory of memory *per se*, yet the past is folded into the present as "residue", or "relics". Also, in German the terms *Erfahrung-Erwartung-Erlebnis* convey something of the semantic shift involved to the categories of the "space of experience" and the "horizon of expectation".

[45] *ibid,* p. 257.

overlap and mutually impregnate one another."[46] Expectations are not so much a question of what is temporally exceeded by the present, as an openness toward a possible future. As such, the present contains within itself in "present-centeredness", a disposition towards the future, as well as a reflection of the past, although these two temporal categories do not conflate one to the other.

Although existing upon different temporal layers, the future always remains tied to the past, albeit with "remains" that can never be fully incorporated. Similarly, in opening a new horizon of expectation, the progression is tempered by non-contemporaneous elements occurring simultaneously. "The one process of time became a dynamic of a coexisting plurality of times."[47] In proposing the categories of the "space of experience" and the "horizon of expectation", Koselleck wishes to emphasize the asymmetry and a-chronic character inherent in these categories, which does not arise in the categories of past-present-future; where from a point, the infinity of the past time and the infinity of future time stretches out indeterminately.

In summary, Koselleck's methodology of historicity attempts to acknowledge the singularity of individual points-of-view within a totality of a temporal horizon. Furthermore, "time is no longer simply the medium in which all histories take place; it gains an historical quality. Consequently, history no longer occurs in, but through time. Time becomes a dynamic and historical force in its own right."[48] As temporal markers or borders, the past, present and future coalesce into one another in an immanent mode of experience in exponentially multiple ways. Various eras or epochs have their own singular way of locating themselves with respect to temporal structures. Time itself changes and moves.

Poulantzas' Space-Time Matrix

Of course, if the time of history is considered to be multiply layered, then the problem arises of how to account for the synchronicity of events. Certain cross-sections through the layers, as it were, are needed in order to describe the "now", or a specific time that is concurrent with a linear meted-out time framework, that time that is known as years, decennia, and centuries of time.

Nicos Poulantzas, in *State, Power, Socialism*,[49] was primarily concerned with problematizing the position of the Nation State in capitalist development; how the political articulates itself in economic production, an

---

[46] *ibid*, p. 262.
[47] *ibid*, p. 269.
[48] *ibid*, p. 237.
[49] Poulantzas, Nicos, *State, Power, Socialism*, translated by Patrick Camiller, introduction by Stuart Hall (London and New York: Verso, 2000).

area remarkably absent in the economic theory of Marx.[50] Of course, Marx based his critique of capitalism on the social structures constituted by economic relations; these structures, albeit "political", are not necessarily governmental forms, such as the nation state in terms of the twentieth century. Many European countries had some form of socialist government after the Second World War. Precisely this condition, along with the post-1968 philosophers and intellectuals, many of whom were the inheritors of Lefebvre's work on the State mode of production and Althusser's Marxist critiques, were in turn carefully treated in a discussion in Poulantzas' *State, Power, Socialism.*

Specifically, the shake-up caused by Foucault's work on the manifestations of power in knowledge production and socio-political institutions, prompted Poulantzas to extend the argument he began in *La crise de l'Etat* of 1976.[51] In the forms of nation states that have arisen within the capitalist development, several phases can be characterized. Specifically in this section, the characterization of the development within a territory (space) and a tradition (time) were brought into a theoretical substructure that Poulantzas called, "the space-time matrix". The nation state and economics are not two separate domains, but rather Poulantzas asked: "How does one intervene in the other and indeed, how does the border become renegotiated over time, the contours of which are constantly changing?"[52] The state in economic theory cannot be assumed to be an equivalence of "nation" with all the inherent problematic of borders and internal circulation of capital globally. Yet the concept that Poulantzas developed of a spatial-temporal matrix, positions the state within the nexus of capital-social institutions and forms of governance, including the revolutions thereof.

Poulantzas provides a useful theoretical concept of the space and time matrices. A cautionary note, however: these matrices are not to be thought of as some sort of "diagram", mapped onto national territories or historical periods. Rather, these spatio-temporal matrices are coincident with the event of their occurrence. As such, they arise or are spontaneously created at once as the presupposition of a certain socio-economic frame, as well as the conditions of possibility for further (capitalist) development. Indeed, the matrix is immanent to the human condition.

Poulantzas describes the spatial matrix as territory, and the temporal matrix as historicity and tradition. Both are continuous, even though they are "uneven" and heterogeneous. Certain aggregates or densities of the

---

[50] Poulantzas, *op cit*, p. 93: "In fact, we have to recognize that there is no Marxist theory of the nation".
[51] Poulantzas, Nicos, *La crise de l'Etat* (Paris: Presses Universitaires de France, 1976); and Poulantzas, *State, Power, Socialism*, pp. 204-5.
[52] Poulantzas, *State, Power, Socialism*, p. 166.

spatio-temporal matrix come together to form communities, tribes, and nation states. Obviously, the nation state contains certain commonalities of language, borders, and traditions. However, as history itself proves, these borders are often less than hard and fast. Events are not inscribed into the matrix; rather, time as history and space as territory emerge from the continuous flow of the spatio-temporal matrix.

Furthermore, the notions of "interior" and "exterior", or "inter-tribal" and "barbarian", or "us" and "them" have no precise "demarcation". "Indeed, why and how are such limits assigned to a particular field, which becomes the site of the problem of unification?"[53] On one side, the borders of a nation state are somewhat "fuzzy" in terms of the fact that the border contains an "inside" of heterogeneous elements – different languages, cultures, and historical traditions. On the other side, the advent of global capitalism does not, according to Poulantzas, eradicate the differences, or even-out the field. So, even though in Marxist terms in contemporary societies the unevenness of capitalist development still holds, the border between the nation states which are "developed" and those which are "undeveloped", or slower to develop, is no longer a category that can be differentiated as an absolute limit. For indeed, each nation state, or even contested territory, lying outside the periphery of what could constitute a *de jure* nation state, is an intimately interdependent element in the whole spatio-temporal matrix.

In fact, the matrix could be seen as a substratum for the Marxian notion of the superstructure, embodying the various mutations in capitalist development, or indeed any socio-economic development that has historically occurred, or will mutate into some other form, for example, Chinese free market communism, European liberal economic welfare states, or republican totalitarianisms. We must not forget the "unholy alliances" either, multiplicities of diplomatic strong-arming that form interwoven "forces" that effect change outside of traditionally held frontiers. So even though Poulantzas concedes that any given nation state is not reducible to the substratum of a spatio-temporal matrix, as a socio-economic entity, the State intercedes in the material conditions of the matrix, as well as conceivably, like other entities, attempts to control or even dominate the flow of power within the field, and map out its sphere of influence in ever wider terrains or domains.

First, let us examine Poulantzas' notion of the spatial matrix as territory apart from temporal or historical considerations. Obviously, this notion of space is not the same as a phenomenological, experiential, or physical space. Further, this concept is much more pervasive than the Marxist notion of the expansion and consumption of space through economic

[53] Poulantzas, *ibid*, p. 96.

exploitation. Rather, the spatial matrix is part of the substratum, and indeed a condition of all possibility of borders, territory, transportation infrastructure, inter-communications and international relations. Yet the changes that the history of human settlement has undergone are not mere elaborations upon a spatial theme. Territories shift and evolve – and even die – indicating not only variation, but discontinuity. Yet, "the transformation of these mechanisms", according to Poulantzas, "is woven into a more intricate web. We are not dealing with different modes of organization, appropriation and consumption of a "space" that presupposes an intrinsic nature, nor with different trajectories and structurings of one and the same space."[54] Now admittedly, Poulantzas distinguishes between territories pre-capitalist and capitalist, albeit without the somewhat overly facile differentiations of "territorialization" and "deterritorialization" of Deleuze and Guattari.

Although more ancient societies were organized in Western Europe around a feudal chief with his accompanying lands, or a city-state whose power was turned inward toward a religious or secular power often from the same family, these organizational forms were poly-centric, the forces of power centered upon the agora, land house, or church square. This pre-capitalist polycentric condition Poulantzas describes as a "space that is continuous, homogeneous, symmetrical, reversible, and open".[55] Although these spaces form some similarities with the voracious and expanding space of capitalism that "produces space", Poulantzas differentiates these spaces as simply "spread out in a homogeneous field, which, while exhibiting certain delimitations, knows no enclosure in the modern sense of the term."[56]

Although these social economies where decidedly taking on other forms, one could argue that these forms were heterogeneous and in a vast system of relations as centers or city states, even if those relations were often one of war, or aggressive expansion. In contrast to a definitive determination of the pre-capitalist political economies as somehow evenly spread out in a field, these can be seen to be just another instant of the temporal matrix wherein labor, capital, and surplus value are alternatively expressed, and indeed expressed in a way which can be found in contemporary history without calling these political economies explicitly "pre-capitalist".

Yet Poulantzas himself, following Marx, considers the homogeneous field as the precondition of organized capital, and therefore an altogether different manifestation, one that is "serial, fractured, parceled, cellular and irreversible space".[57] Admittedly, Poulantzas in following Marxist theory so closely here, must indeed make this demarcation in the

---

[54] *ibid*, p. 100.
[55] *ibid*, p. 101 (emphasis in the original).
[56] *ibid*.
[57] *ibid*, p. 103.

economic field, the only real difference to the present circumstance is the issue of the border. Even though the capitalist space matrix is composed of various interrupted, partial and fractured components, Poulantzas argues that "it has no end". [58] Indeed, as has been repeatedly argued, capitalism "produces space"; furthermore absolutely needs to produce more markets, cheaper labor, and less expensive materials in order to continue creating surplus value. However, the fact that contemporary economies are inextricably linked does not constitute a real difference in its heterogeneity, or distinction from pre-capitalist societies. Although these economies have not "reached their end", they too were segmented, expansive of territory when possible, and in fact driving towards the end of the earth as a political impulse. In short, the spatial matrix can be seen to be historically continuous without simply reducing the temporal to a standard, linear measurement.

Naturally, when the spatial matrix is seen not as a specific territory with explicit permanently held borders, but an unfolding of relations, of an "inside" that is porous, containing disparate languages, cultures, and traditions, then how does a nation state construct a unity, a juridical entity? Of course, the impulse to expand exponentially effectively will mean that the "other" is interiorized into the territory of the nation state. Consequently, any nation state is confronted not only with a possible gain or loss of territory, but also eventually the "enemy within", which is the place from which genocide is born,[59] attempting to eradicate the "outside" from the enclosure. But most importantly, the nation state is involved in the very construct of the spatial matrix, crossing frontiers, gathering various social "monads"[60] of institutions, corporations, armies, etc. into a not always harmonious homogeneous unity, but rather into an aggregate that can be directed and ruled. Indeed, the very fragmentation of these various social monads ensure that no single entity will gather enough force to threaten the act of creating unity – which is another word for the political and economic power to create a means to act in the spatio-temporal matrix. Indeed, as Poulantzas points out, the "key function of every dominant ideology [is] namely that of cementing together the social formation under the aegis of the dominant class."[61]

Second, although Poulantzas puts forward his concept of the spatio-temporal matrices as a specific manifestation of capitalism, the division of labor, and the estrangement of the worker from the means of production, his notion of the temporal matrix becomes increasingly important in understanding the place of historical determinism and the Marxist ideology of "progress". Time, instead of being ruled by the natural movement of the

---

[58] *ibid.*
[59] *ibid*, p. 107 and p. 114.
[60] *ibid*, pp. 64-65.
[61] *ibid*, p. 88.

planetary bodies, the gentle change of seasons and the course of night becoming day - time becomes repetitive, serial, and linear. In fact, time becomes manifest in the production line,[62] the quantifiable measure of the necessary time of production, and as such, a key component in the estrangement of the worker from his labor. In the space of the commodity circulation as well as the insertion of workers into variable labor tasks, the temporal matrix interpreted as historicity and tradition is not to be thought of as some kind of periodization, or historical series of events, or the passing review of "worldviews". Specifically, Poulantzas wishes to conceptualize a temporal matrix that would include the slowing-down and speeding up of economic development, the unification process that occurs in order to define that somewhat consistent state of affairs known as the nation state, as well as the unending nature of change in the spatio-temporal matrices.

In spite of Poulantzas objecting to the characterization of historical time as periodization, he does make distinctions between the temporal matrices of antiquity, medieval feudalism, and modern industrialization. Correctly, time in antiquity was cyclical, specifically the turning of the harmonic spheres of the heavenly bodies. Undoubtedly, this time was also punctuated by events such as wars, the ascension of various leaders, and repeated natural disasters. Yet, even though multiple times could be seen in the spheres of agriculture, politics, and warfare, they still make up a single "homogeneous temporal continuum".[63] The cyclical time of antiquity was governed by fate, by chance - the gods not being predictable. Yet the revolution of the antique temporal matrix was a chronology that continuously searched for the origins, returning time and time again to recover the present. Indeed, Plato put forward a theory of knowledge known as *anamnesis*, or the remembering of truth innately embedded in the soul of man. In antiquity, the "present" was always present with the origin and with the fates – present in one singular circular temporal matrix.

On the other hand, in medieval times, although sharing commonalities with antiquity, time in Western Europe at least, could be said to "stall", or to slow down, and begin to center on a system of agrarian economies around a particular city-state, or feudal kingdom. Thus, the temporal matrix of medieval time was rhythmically meted out in accordance with the agrarian passing of seasons and the appropriate labor therefore. Time was not an eternal recurrence in the sense of antiquity; rather, an eternity. Specifically, the Christian orthodoxy ruled time. The "present" was always oriented on the one hand, to the original sin, and on the other hand, to the time of divine eternity. "History", argues Poulantzas, "now has a beginning

---

[62] *ibid*, pp. 64-65.
[63] *ibid*, p.108.

and an end, located between the Creation and Judgment."[64] Yet the "before" and "after" of continuous time becomes the ever-present "present" in the temporal matrix of Medieval Christendom.

Both these matrices are at once folded into the pre-capitalist historicity in terms of tradition. Nevertheless, the industrialization of Europe heralds in a completely new temporal matrix: time is oriented to the future, towards Enlightenment, towards utopian progress, towards increased productivity. For Poulantzas, the capitalist temporal matrix occurs when time splits apart as it were, into multiply divided, albeit cumulative time. Most importantly, time is measureable – the telegraph, the railroad, the portable clock, machination of production, and electric lighting all change the way that not only labor, but also society is organized. "This temporal matrix for the first time marks out the particular temporalities as differential temporalities – that is to say, as rhythmical and metrical variations of a serial segment, irreversible and cumulative time."[65] Obviously, in this historical period, man is delineated as a punctual mark on the evolutionary temporal scheme, evolving and adapting. Time does not loop back to the beginning as in antiquity, or is it suspended in the divine grip of the "now". Time is forever open to the future as progressive teleology.

Subsequently, the unification of the temporal matrix becomes problematic if time cannot be integrated into a continuum. For Poulantzas, the emerging nation state in modernity jumps into the gap, and assigns a temporal matrix to the "present", providing "the standard of measurement, or in other words, the frame of reference of the variations of particular temporalities," unifying temporal events, and determining the direction of their unfolding. This unifying force through the disparate temporal planes, "pushes the back" of time at it were, where the "present" is a mere precondition to a possible futurity.

The nation state gathers time and assigns an aim without a final end. Time becomes a mere segment upon a measurable directed path. Indeed, the nation state becomes the relationship that also ties together the spatial and temporal matrices, "between history and territory [in fact], the modern unity thereby becomes historicity of a territory and territorialization of a history."[66] This unification in modern capitalism, then, is brought into effect through the means of power. Of course, power need not be ***force***. However, the danger always exists in the unifying of the disparate spatio-temporal matrices, in tying together various languages, traditions, and peoples, in "separation and division in order to unify; parceling out in order to structure; atomization in order to encompass; segmentation in order to totalize; closure in order to

---

[64] *ibid*, p.109.
[65] *ibid*, p.110.
[66] *ibid*, p.114.

homogenize; and individualization in order to obliterate differences and otherness."[67] When power becomes totalitarian, leveling out space into the segmented, then time, in fact, freezes. Then, the "present" becomes a never-ending suffering without end.

To conclude, Poulantzas with his notion of the spatio-temporal matrix, provides an important theoretical tool in order to understand history not just in terms of serial events or periods; and to understand space not just in terms of the ever-shifting boundaries between territories, but also as a field in which various components, interrupted and fragmented, structure or develop their relations. Of course, given that time is layered, and space is multiple, the problem arises of how to account for the synchronicity of events. For Poulantzas, the nation state was the unifying force, albeit with the ever-present danger of the differences being evened out and becoming instead an expression of totalitarianism. In trying to control the disparities, variation and diversity are not tolerated. Nonetheless, the relationship is made through means of power – and power that is not necessarily violence. And it is to Foucault that we now will turn, to his extensive discourse of the theme of power, in order to account for how a spatio-temporal field expresses, expands, and extends.

Tying together the Layers: Foucault's Conception of Power

On questions of power and space, Foucault immediately springs to mind. Although he makes protestations that his philosophical inquiry was not about the analysis of power, or its mechanisms,[68] certainly the trope of power reappears prolifically throughout his oeuvre. Emphatically, for Foucault, his question has been the human subject and its constitution as a subject, and indeed as an object for itself. Thus, power is not properly speaking the theme of his research; however, power (or what Foucault will come to call bio-power) returns again and again as a constituent question, albeit in a very complex system of relations. Foucault does not provide a "theory of power" because, as he argues the possibility of a theory of power presupposes an object upon which power is exercised or resisted. Yet precisely the step before – how man becomes an objectified or reified or conceptualized subject – is the way of thinking of power that encompasses Foucault's genealogical method.

Tracing these pathways or constitutions of power historically and in its diverse forms, Foucault extends the definition of power beyond its mere

---

[67] *ibid*, p.106.

[68] Foucault, Michel, "Subject and Power" in Dreyfus, Hubert L. and Rabinow, Paul, *Michel Foucault: Beyond Structuralism and Hermeneutics* (Brighton, UK: Harvester Press, 1982), pp. 208-226.

"pathological forms".[69] Although the strategies of power revolve around the institutional exercise of power, and resistances to such on the part of individuals – and indeed constituting themselves as such through these acts and resistances – Foucault's purpose is not to critique the institutions as much as to reveal the workings of various "techniques" of power.

This form of power applies itself to immediate everyday life which categorizes the individual, marks him by his own individuality, attaches him to his own identity, imposes a law of truth on him which he must recognize and which others have to recognize in him. It is a form of power which makes individuals subjects. For Foucault, there are two meanings of the word subject: subject to someone else by control and dependence, and tied to one's own identity by a conscience or self-knowledge. Both meanings suggest a form of power which subjugates and makes subject *to*.[70]

Consequently, Foucault elaborates three forms of power: domination, exploitation, and subjection. Each of these three indeed is operable, inter-twining in their techniques, and forms not only a kind of "positive" form of power, but also the very resistances to domination and to exploitation. However, escape is not possible; which is not to say that these forms or strategies of power are homogeneous, rather form or constitute a kind of force field where the human as subject is determined qua being human, and the mechanisms of which form all other economic, social, and ideological structures.

In short, Foucault argues, "the mechanism of subjection (i.e. subject-forming and subject-resisting) cannot be studied outside their relation to the mechanisms of exploitation and domination."[71] Specifically in the matrix of power that constitutes the state, the structure is "both an individualizing and totalizing form of power."[72] As such, the individual is subjected simultaneously to acts of individualization and totalization on the part of modern power structures. Resistance would entail constructing new forms of subjectivity, according to Foucault, yet "subjection" remains the constitutive power of the human subject intertwined with other forms or mechanisms of power. The exercise of power cannot be avoided; rather in resistance, power relations can become less exploitative or repressive. Therefore, Foucault states:

> Power relations, relationships of communication, objective capacities should not therefore be confused. This is not to say that there is a question of three separate domains. Nor that there is on

[69] Foucault, "Subject and Power", *op cit,* p. 209.
[70] Foucault, *ibid,* p. 212.
[71] *ibid,* p. 213.
[72] *ibid.*

> one hand the field of things, of perfected technique, work and the transformation of the real; on the other hand that of signs, communication, reciprocity, and the production of meaning; finally that of the domination of the means of constraint, or inequality and the action of men upon other men. It is a question of three types of relationships which in fact always overlap one another, support one another reciprocally, and use each other mutually as means to an end."[73]

Decidedly, the three forms of power – domination, exploitation, and subjection – exist in a very complex system of relations in a spatio-temporal field.

Foucault and the Turn to the Body

Foucault, in his examination of the constitution of the human subject, turns to the body itself as a node in a relational field. What one might surmise in the discourse on the body is that this move entailed a radical subjectivity; since Foucault's self-professed (confessed) philosophical problem is subjectification as part of the exploitation/domination/subjection triad. Escaping the modernist subject/object divide, Foucault does not see the subject in terms of "subjectivity", intentionality or psychology, but rather as in a nexus of social forces, in a network of the production of knowledge and the institution of social practices that in themselves are the interplay of various strategies of power. Furthermore, these social practices and "wills to truth" are not transcendentally directed or developing or unfolding along some historical deterministic principle, progressing towards some ultimate *dénouement*. Foucault, with his genealogical method,[74] merely traces the strategies and effects of power in a network of relations.[75] Each strategy of power exists in "a common historical matrix".[76]

Most importantly, in Foucault's turn to the body, the body is no longer the point from which all subjectivity occurs, or the site of the

---

[73] *ibid*, p. 218.

[74] Foucault, Michel, *The Care of the Self: The History of Sexuality*, volume 3, translated by Robert Hurley (New York: Vintage, 1988), p. 262. In the essay, "On the Genealogy of Ethics", Foucault outlines the three domains of genealogy: "First, a historical ontology of ourselves in relation to truth through which we constitute ourselves as subjects of knowledge; second, a historical ontology of ourselves in relation to a field of power through which we constitute ourselves as subjects acting on others; third, a historical ontology in relation to ethics through which we constitute ourselves as moral agents."

[75] Foucault, Michel, *Discipline and Punish: The Birth of the Prison* (New York: Vintage Books, 1979), p. 25: "the body is also directly involved in a political field. ... Power relations have an immediate hold upon it; they invest it, mark it, train it, torture it, force it to carry out tasks, to perform ceremonies, to emit signs". And further, on page 26: "[T]he body becomes a useful force only if it is both a productive body and a subjected body".

[76] Dreyfus and Rabinow, *op cit,* p. 160.

ontological gap; rather, it is the nexus of intimate and social practices or a node in a network of power relations. Institutions are sites of production, or technologies of power that articulate power relations. In the radical proposition that knowledge and power are embedded in the generative site of the body, Foucault critiques not only modernist conceptions of subjectivity, but also the objective and a-historical pretensions to a universal truth. The matter is indeed much more complex, and genealogy is the method by which the strands of power's articulation can be shown in their operative nature.[77] Yet furthermore, the body as generative site is not to be understood as a subject with consciousness and will as defined by modernism. "We must", Foucault writes, "rid ourselves of the constituting subject, rid ourselves of the subject itself, which is to say arrive at an analysis which can account for the subject within an historical account."[78]

Indeed, as Foucault so convincingly showed in *The Order of Things*, the subject itself is constituted, and indeed, constituted for itself as a subject. Dreyfus and Rabinow point out that, "the modern individual – objectified, analyzed, fixed – is a historical achievement. There is no universal person on whom power has performed its operations and knowledge, its inquiries. Rather, the individual is the effect and object of a certain crossing power and knowledge. He is the product of the complex strategic developments in the field of power and the multiple developments in the human sciences."[79] In contrast to the conception of "subject" as opposed to "object" that is inherited from modernism, with all the constituent problems of how this human subject knows the world and how the subject is related to the world, Foucault articulates the conception of "bio-power". Foucault defines "bio-power" quite simply as the forms of life in a field of power.

Furthermore, these articulations of the technologies of power can be discursive and/or non-discursive. For Foucault, in the triad of exploitation/domination/subjection, there is also the opposition or resistance to exploitation and domination. In the body, resistances to the mechanisms of power also entail the refusal to comply with institutional techniques of subject-forming. However, the extent to which a pure resistance is possible, or even desirable, is not a matter of "equal and opposite". Rather, power and its resistance should be seen as a field of forces, dynamically interchanging and localized in the body's potential and capacity to act. "Foucault holds that power needs resistance as one of its fundamental conditions of operation. It is through the articulation of points of resistance that power spreads through the

---

[77] Foucault, *Discipline and Punish*, p. 27: "we should abandon a whole tradition that allows us to imagine that knowledge can exist only where the power relations are suspended and that knowledge can develop only outside its injunctions, its demands and its interests."

[78] Foucault, "Truth and Power", in: Gordon, C. (ed.), *Power/Knowledge: Selected Interviews and Other Writings 1972-1977* (Brighton: Harvester Press, 1980), p. 117.

[79] Dreyfus and Rabinow, *Foucault,* p. 159-160.

social field. But it is also, of course, through resistance that power is disrupted. Resistance is both an element of the functioning of power, and a source of its perpetual disorder."[80]

The Panopticon is not Hidden

The panopticon, the hospital, and the church confessional booth are merely the ritualistic sites where power is exercised. They do not cause power to be executed, rather a conflation or localization, or the utilization of power. The panopticon is not hidden. As Dreyfus and Rabinow point out: "it should be emphasized that prisons are only one example among many others of this technology of discipline, surveillance, and punishment [being only one of the] clearly articulated expressions of more generalized practices of disciplining both individuals and populations."[81] As such, the physical architecture embodying these technologies of power is not the object of study; rather, the procedures of power as a technique for discipline. The panopticon is mere a non-discursive player, so to speak, in the articulations of power.[82]

The most important historical turn that makes these procedures possible is the consideration of the body as object, as subject, as a focal point of study, i.e. "bio-power", and as something that needs to be disciplined, exploited, controlled, or formed (educated). In order to implement these strategies, the body must be objectified, atomized, and then pacified. This strategy needs to be incessant in order to maintain control over the body, and space becomes complicit in this disciplinary organization.

In fact, the turn to bio-power that Foucault identifies, is the very one where precisely the body in the penal system would need to be held, to be incarcerated and to be "reformed" (as opposed to exiled or summarily executed), is in itself a change in the consideration of the body as subject. Yet, this turn is not to say that certain forms of architecture – prisons, hospitals, factories, monasteries, educational institutions – *cause* the disciplinary strategy; rather, it is a constituent procedure in the constitution of man *qua* bio-power as something that needs to be controlled. As such, these spatial forms constitute a non-discursive grid that acts in conjunction with other power strategies, such as knowledge production, jurisprudence, and social practices, to organize the disciplinary action of subjection.[83]

---

[80] Dreyfus and Rabinow, *ibid*, p. 147.

[81] *ibid*, p. 153.

[82] Foucault, *Discipline and Punish*, p. 205. "The Panopticon," Foucault argues, "must be understood as a generalizable model of functioning: a way of defining power relations in terms of everyday life of men."

[83] *cf.* Allen, Barry, "Foucault and Modern Political Philosophy" in: Moss, Jeremy (ed.), *The Later Foucault* (London: Sage, 1998), p. 168. "In conjunction with statistical techniques developed since the nineteenth century, the new information and communications technologies of the computer age enable a system of disciplinary surveillance not unlike that of Bentham's

"Discipline", Foucault argues, precisely "'makes' individuals; it is the specific technique of a power that regards individuals both as objects and instruments of its exercise."[84] The body, and not the building *per se*, is the locus of techniques of power. "Rather, the individual is the effect and object of a certain crossing of power and knowledge. He is the product of the complex stategic developments in the field of power and the multiple developments in the human sciences."[85] As Dreyfus and Rabinow explain:

> Hierarchical observation is a key element in the examination. The goal is to make surveillance an integral part of production and control. The act of looking over and being looked over will be a central means by which individuals are linked together in a disciplinary space. The control of bodies depends on an optics of power. The first model of this control through surveillance, efficiency through the gaze, order through spatial structure, was the military camp. Here total organization and observation were possible. The functions performed here were limited, but the model worked; and later it spread to the construction of grand urban schemes, working class housing projects, prisons, schools, and so forth. On a broad scale, the model of the military camp provided control through hierarchy and observation. But it first had to be refined in other settings....The means for accomplishing this were in part architectural.[86]

This model of control became almost ubiquitous and universally applicable. In the age of industrial production, the factory floor also became a matter of surveillance and control of its workers in order to assure maximum productivity and passive subjection to the minute role that each worker must play within the whole manufacturing process. Each act of labor was quantified, disciplinary measures needed to be overseen, and power

---

Panopticon, but without its clumsy architectural constraints. Technology-assisted authority can be exercised remotely at multiple sites and at a fraction of the cost of physical presence, vastly enhancing administrative capacity (private and public) at the expense of individuals and unorganized sectors of society. The now routine use of such technology evinces a trend towards an actuarial conception of social control. Relying on high-speed sorting and analysis, individual records can be compared with statistically constructed normative "profiles" and pre-emptive measures taken on the basis of probable deviancy, the would-be delinquents systemically "pre-detected" before they have the opportunity to disobey. Under these conditions, civil law comes to be identified with the authoritative redistribution of social risk, and the administration of justice with a mechanism of social control." *cf.* Gandy, O. H. Jr., *The Panoptic Sort: A Political Economy of Personal Information* (Boulder CO: Westview Press, 1993).

[84] Foucault, *Discipline and Punish*, p. 170.

[85] Dreyfus and Rabinow, *op cit*, p. 160.

[86] *ibid*, p. 156-7.

distributed hierarchically. Although in *Discipline and Punish: The Birth of the Prison*, Foucault argues that the body is the site and object of control for the seventeenth century, the body under constant surveillance, incarceration and "correction" becomes the site of power through discipline from the eighteenth century onwards. The body is detained and corrected, as opposed to publicly or spectacularly tortured and executed.

This new politics of the body, where the body as object of power is caught in a system of power relations, only indicates a much larger network of relations at work. Life itself – at work, at church, at home – became partitioned in a strict timetable, under surveillance of various institutions of disciplinary control. The most important vision of control is the metaphysics of transcendence: "God sees everything", the omniscience of the transcendent, where power is exerted through the mechanism of control of the all-seeing eye of God. The examples provided by Foucault – mental hospitals, education institutions, and sexual pedagogy – are mere microcosms of power, "distributing individuals, fixing them in space, classifying them, extracting from them the maximum in time and forces, training their bodies, coding their continuous behavior, maintaining them in perfect visibility, forming around them an apparatus of observation, registration and recording, constituting on them a body of knowledge that is accumulated and centralized."[87] There is nothing outside of power - either the act, or the resistance to the act.

Nevertheless, these "non-discursive" and discursive mechanisms of power analyzed by Foucault are in the end in the network of relations that constitute the subject as "man", as modernist "subject", and as a body politic. As Foucault reminds us though:

> [I]t should not be forgotten that there existed at the same period a technique for constituting individuals as correlative elements of power and knowledge. The individual is no doubt the fictitious atom of an "ideological" representation of society; but he is also a reality fabricated by the specific technology of power that [he has] called "discipline". We must cease once and for all to describe the effects of power in negative terms: it "excludes", it "represses", it "censors", it "abstracts", it "masks", it "conceals". In fact, power produces; it produces reality; it produces domains of objects and rituals of truth.[88]

Importantly, Foucault in the end escaped the binary relationship entailed in the axis: power as force, or the abuse of authority, and the mere resistance to this power. Power becomes a productive or generative concept in the ability

[87] Foucault, *Discipline and Punish*, p. 231.
[88] Foucault, *ibid*, p. 194.

to act for oneself, and more specifically to "know oneself". Of course, "power relations are extremely widespread in human relationships...The analysis of power relations is an extremely complex area; one sometimes encounters what may be called situations or states of domination in which the power relations, instead of being mobile, allowing the various participants to adopt strategies modifying them, remain blocked, frozen."[89] However, breaking through the obstacles becomes more than a question of resistance to the control of others. Power, albeit inescapable, also produces.

Technologies of the Self

Towards the end of his life, Foucault changed the scale of his inquiry. On one hand, his final seminars at the *Collège de France* encompassed the scope of global politics, economic structures and supra-national governance. On the other hand, given his emphasis on a critique of subjectivity throughout his career, Foucault turned to the most intimate constitution of the human subject – "care" or control over, or responsibility "to know oneself". The "cultivation of the self" as defined by the ancient Greeks, assumed that the goal of life is to be happy, and this happiness is attained through the search for wisdom (*philos-sophos*), and for which certain "techniques" exist as passed down from teacher to student. These are the "techniques of the self".

Foucault takes up an exegesis of the Platonic dialogue *Alcibiades*.[90] In this rich text, Socrates has waited patiently for the beautiful Alcibiades to come of age. He then accosts him in his usual manner, with questions: "what do you know?" he demands. Alcibiades then proceeds to tell him of his athletic prowess, his great wealth, his fortunate birth, and his beauty that is desired by all who see him. As one who will eventually take up a position of power in the society, a leader must know who is the ruled, and who is the ruler. "Anyone who exercises power has to place himself in a field of complex relations where he occupies a transition point."[91] In short, a city-state could only prosper and her inhabitants be secure and content if the leaders were "philosophers", capable of the responsibility for the "care of the self", and their decisions just, measured, and good. Of great import was the manner in which the leaders exercised their power and authority, and indeed knew the limits of their power. Yet firstly, "anyone who governs...must attend to himself, guide his own soul, and establish his own *ethos*."[92]

---

[89] Foucault, Michel, "The Ethics of the Concern for the Self as a Practice of Freedom", in *Ethics: Subjectivity and Truth,* edited by Paul Rabinow, (New York: The New Press, 1997), pp. 281-301.
[90] Plato, *Alcibiades,* in Greek with comments by and edited by Nicholas Denyer (Cambridge: Cambridge UP, 2001).
[91] Foucault, Michel, *The Care of the Self: The History of Sexuality*, Volume 3, translated by Robert Hurley (New York: Vintage, 1988), p. 88.
[92] Foucault, *ibid*, p. 89.

To this end, Socrates begins to groom Alcibiades. But Alcibiades is ignorant and intellectually indolent. He will need to know many things if he will wield power well. "What", demands Socrates, "does Alcibiades intend to say to the Athenian Assembly when he comes of age?" At twenty, Alcibiades assumes he will enter the political arena by storm. Alcibiades confuses his wealth, beauty, and physical courage with the wisdom of knowing how to use these things rightly.[93] Ignorant is Alcibiades, and not in the Socratic sense, where one knows one is ignorant, knows the limits of understanding. The most important thing to know is that a man is nothing other than a soul (τὴν ψυχήν).[94] However, Alcibiades objects, why should I know when no one else knows, including my esteemed colleagues in the Assembly, as well as our enemies? Yet they think they know without the effort of finding out, either for themselves, or from others. Alcibiades plans instead rely upon his birth and his ability to intimidate and to subjugate. However, Socrates sarcastically reminds him, "it is curiously illogical that slave owners should despise their slaves for cowardice, yet not wish them to be brave enough to rebel."[95] Is he not brave, he, who can bully slaves?

"What then do you think will happen to men without virtue, who have no care for themselves?", asks Socrates.[96] Of course, the injunctive "know yourself" (γνῶθι σαυτόν) was prominently inscribed over the temple of Apollo at Delphi. But this fundamental principle meant more than the mere imperative to know, to have *techné*, to discern the good and beneficial, to exercise power justly. Instead, man should know most of all that he is a soul, and that he is responsible for its "care". The skills necessary to take care of oneself, Alcibiades confuses with taking care of his possessions. Yet, the care for essential things is different than caring for mere possessions belonging to a man. Finding Alcidiades to be an incorrigible student, Socrates persists. Socrates drives home the final question: "Could we ever know by what art (*techné*) we could care for ourselves, if we are completely ignorant of what we ourselves are?" (τίς τέχνη βελίω ποιεῖ ἄνθρωπον, ἆρ' ἄν ποτε γνοῖμεν ἀγνοοῦντες τι ποτ' ἐσμὲν αὐτοι).[97]

Yet the only way we can come to know what we ourselves are, is through the mirroring of our soul by another soul. This mirroring is done by a teacher who helps the student "remember" (*anamnesis*), who he is. But also, most importantly, we come to know who we are through our "other half", through another soul who loves and is loved, through "itself by itself" (αὐτο

[93] Plato, *Alcibiades, op cit,* p. 243-245. (134e8-135b6). See also: Foucault, Michel, "The Hermeneutic of the Subject", in *Ethics: Subjectivity and Truth, op cit,* p. 87.
[94] Plato, *Alcibiades, op cit,* p. 155. (130c3).
[95] *ibid,* p. 181. (122a8).
[96] *ibid,* p. 189. (123d8-e2).
[97] *ibid,* p. 210. (128e10-11).

τὸ αὐτό).[98] The most superior way to know the soul is in the mirror reflected by the gods. Alcibiades must come to know himself, and care for his soul if he is to lead wisely and to judge fairly. Ignorance is dangerous, for if he does not become virtuous, wise, and just, he will not only waste his soul, but also endanger the lives of his community. Only a man truly capable of caring for himself, governing himself, can govern or care for others. This viewpoint that he who governs must know how to care for himself, is a frequent pedagogical theme in ancient Greece. A wise man is a happy man and "he must attend to himself, guide his own soul, establish his own *ethos*."[99] This *ethos* is not an ethics, and even less a morality; rather, it is the ultimate taking responsibility for one's own soul.

Power as an Intimate Ethos

Foucault was the consummate philosopher to the end, attending to the work of the soul, "taking care" of the self. Never complacent, the philosopher must question, must attend to, must finally come to govern himself, and to know his own soul.[100] As such, these last essays and lectures at the *Collège de France* constituted a completion of his life-long studies of subjectivity: how a subject is constituted in a network of power relations, how a subject is subjugated, and finally how a subject comes to know itself. "The guiding thread that seems the most useful for this inquiry is constituted by what one might call the 'techniques of the self', which is to say, the procedures, which no doubt exist in every civilization, suggested or prescribed to individuals in order to determine their identity, maintain it, or transform it in terms of a certain number of ends, through relations of self-mastery or self-knowledge." [101] Indeed, in the end, power became more than the opposition of power over, and power to resist control. Power became generative and productive. Power is knowledge of the soul.

---

[98] *ibid*, p. 213-217. (129b1 and 130d5).

[99] Foucault, *The Care of the Self, op cit,* p. 89. One is reminded of Heidegger's exegesis of the Heraclitus fragment 119: *ethos anthropoi daimon*, which Heidegger translates as "man dwells, in so far as he is man, in the nearness of god". This exegesis shows that the "attending to", is a close relation to the gods, a reflection of the mirror of the soul in the eyes of god. But also, the *ethos anthropoi daimon* is a statement on the part of Heraclitus that the soul or spirit (*daimon*) in man (*anthropoid*) has its own ethos, its own "home", and the soul must be taken care of, attended to as the hearth burning in the center of the home. Heidegger, Martin, "Letter on Humanism", translation of "*Brief über den Humanismus*", by Frank A. Capuzzi with J. Glenn Gray, published in the volume *Martin Heidegger: Basic Writings*, edited by David Farrell Krell (San Francisco: Harper & Row, 1977), pp. 193-242.

[100] Foucault, Michel, "Technologies of the Self", in *Ethics: Subjectivity and Truth, op cit,* p. 235: "Since we have to take care throughout life, the objective is no longer to get prepared for adult life, or for another life, but to get prepared for a certain complete achievement of life. This achievement is complete at the moment just prior to death."

[101] Foucault, *ibid,* p. 87.

Yet, we have seemingly wandered far from the starting point: how does the spatio-temporal matrix tie together? Foucault avoids the simplistic juridical analysis of power. Instead, he argues:

> rather than looking for the single form, the central point from which all the forms of power would be derived by way of consequence or development, one must first let them stand forth in their multiplicity, their differences, their specificity, their reversibility: study them therefore as relations of force that intersect, interrelate, converge, or, on the contrary, oppose one another or tend to cancel each other out.[102]

Through the relationships of power – the exercise of, resistance to, and production of power, the human subject is constituted. At the nodal point of all historical time and spatial territory, in Foucault's terms, is the human subject. In the end, it is the body that is the spatio-temporal zero-point, as Nicos Poulantzas argues, the "body which unifies space and installs public man within private man: it is a body with no place and no frontiers."[103] Power becomes a productive or generative concept in the ability to act for oneself, and more specifically to "know oneself". Power, as part of the exploitation/domination/subjection triad, ties together historical time. Yet, most importantly, in the end – at the end – power is the most intimate relation of the philosopher to his soul – the power to care for his own soul. As such, power in the nexus of spatio-temporal relations, albeit inescapable, becomes generative and productive.

---

[102] Foucault, Michel, "Society must be Defended", in *Ethics: Subjectivity and Truth, op cit,* p. 59.
[103] Poulantzas, *State, Power, Socialism*, p. 101.

Places Lived in Time*

Mary Lynne Ellis

Through an engagement with the temporality of the subject and a relinquishing of discrete notions of "external" and "internal", Ellis allows for a notion of subjectivity in which the psychic and the social are integral to one another. Although Freud's notion of the unconscious as "timeless" lead to notions of the unconscious as containing raw and primitive forces which are assumed to be ahistorical and asocial in their origins, Ellis carefully examines numerous psychoanalysts, theorists, philosophers, and writers in order to in order to discuss how a theorizing which prioritizes the temporality of the subject allows for more inclusive interpretations of individualities as lived in their socio-historical and cultural specificity. Emphatically, the individual's historicity is contextual; the subject is intrinsically social and linguistic, embedded in a network of human relationships at a particular moment in historical time; we are always already in the world and this is a world of others. Merleau-Ponty, in *Phenomenology of Perception,* theorizes how we are embodied through and by the passage of time, including our anticipation of the future. He argues that subjectivity is always contextual and his theorizing is rooted in his challenge to Cartesian mind-body dualisms, including distinctions between "inner" and "outer", "internal" and "external". He criticizes the claims for universality in psychoanalytic theories with their failure to take into account socio-cultural and socio-historical contexts. Merleau-Ponty's notion of "sedimentation" allows for the acquisition and retention of historical and cultural meanings which inform our ways of being in the world. Ultimately, psychoanalytic attunement to the socio-historical requires an acknowledgement of difference and of change, allowing for multiple meanings of experience, which shift as they are perceived from within different temporal contexts.

This essay discusses how a theorizing which prioritizes the temporality of the subject allows for more inclusive interpretations of individualities as lived in their socio-historical and cultural specificity. In contrast to the classical psychoanalytic focus on the patient's "inner world" (as distinct from the "external world"), the analyst's attentiveness to the socio-historicity of patients in analysis is crucial to his/her understanding of the richness and complexity of patients' experiences (experiences which could not have occurred at any other historical moment) in their multifarious aspects. Such attentiveness also allows for a notion of change: each of our experiences arises from a specific moment in time; they are not fixed and unchanging. This does not mean that it is always appropriate to interpret the socio-historical specificity of an individual's experience directly; rather the orientation of the analyst to this aspect of the material can in itself convey an openness to subjectivity as fluid and changing.

* Places Lived in Time", a revised edition of the chapter originally published in Mary Lynne Ellis, *Time in Practice: Analytical Perspectives on the Times of Our Lives*, published by Karnac Books in 2008, reprinted with kind permission of Karnac Books.

Andre Green, as a contemporary French psychoanalyst, might be expected to consider this aspect of temporality in his book, *The Psychoanalysis of Time.*[1] However, Green considers it unimportant whether fantasies are transmitted genetically or culturally from generation to generation. His theorizing of time emphasizes the value of Freud's notion of the unconscious as "timeless", and of resurrecting Freud's drive theory. Green's lack of interest in the socio-historical is particularly surprising given the accessibility of modern European philosophical thinking in France in which the notion of "historicity" is crucial.

The concept of "historicity" in the hermeneutic and phenomenological traditions in philosophy is very relevant to psychoanalytic practices. "Historicity" denotes an individual's lived experience of their past and present in anticipation of their future possibilities. Furthermore, the individual's historicity is contextual; the subject is intrinsically social and linguistic, embedded in a network of human relationships at a particular moment in historical time. We are always already in the world and this is a world of others. This embeddedness is highlighted by the historian Bourke in his book *Fear.*[2] Assumed to be timeless and universal, the emotion of fear instead "acquires meaning through cultural language and rites."[3]

Initially, a critical consideration of the contributions of some psychoanalysts who, in contrast to Green, explicitly struggle with the question of the relation between the individual and the socio-historical and cultural contexts will be explicated. These psychoanalysts offer radical and courageous challenges to the emphasis on the intra-psychic in mainstream psychoanalytic theorizing. Nevertheless, their work founders on contradictions arising from arguments reliant on the spatial metaphors of "internal" and "external", with the exception of Samuels.[4] Both O'Connor and Ryan[5] point to this problematic in their innovative analysis of the psychoanalytic theorizing of lesbian sexualities.

Through an engagement with the temporality of the subject and a relinquishing of discrete notions of "external" and "internal", we can allow for a notion of subjectivity in which the psychic and the social are integral to one another. The relevance to psychoanalytic practices of the theorizing on time is highlighted through the work of the French phenomenologist Merleau-Ponty and also of the autobiographical work of Eva Hoffman, the contemporary writer and novelist. Hoffman's writing illustrates clearly the

---

[1] Green, A., *Time in Psychoanalysis*, translation by A. Weller, (London: Free Association Books, 2002).
[2] Bourke, J., Fear: *A Cultural History* (London: Virago, 2005).
[3] Bourke, *ibid*, p.7.
[4] Samuels, A., *The Political Psyche* (London and New York: Routledge, 1993).
[5] O'Connor, N. and Ryan, J., *Wild Desires and Mistaken Identities: Lesbianism and Psychoanalysis* (London: Karnac, 2003).

limitations of any theorising which perceives the individual primarily in terms of alinguistic drives, mechanisms or universal complexes.

Gender in Time

The late 1970's, 1980's and early1990's saw the emergence of a range of critiques of traditional psychoanalytic theorizing regarding its failure to attend sufficiently to the relation between gender and the wider socio-historical context. A range of differing and challenging perspectives were developed by psychoanalysts in Britain and the United States, such as Mitchell[6], who draws on Lacanian theory; Dinnerstein[7], who is influenced by Klein; Chodorow[8], Eichenbaum and Orbach[9], whose work is rooted in a critical appreciation of Winnicottian theory, and Benjamin [10], whose theorizing draws on the work of the Frankfurt School. In France, psychoanalysts such as Irigaray[11] and Kristeva[12] were exploring the question of gender against the background of contemporary French philosophy and the rise in popularity of Lacanian psychoanalysis, with its emphasis on the phallus as primary signifier. All of these writers in their radical critiques take the socio-historical and socio-cultural into account. Focusing on the conceptions of temporality which explicitly and/or implicitly underpin their theorising with regard to socio-historical specificity, two examples will be discussed here; one from the Anglo-American feminist discourse (Eichenbaum and Orbach) and one from the French (Kristeva).

Eichenbaum and Orbach's project arose from questions raised by the Women's Liberation movement in Britain and the US in the 1970's, regarding the relation between the personal and the political, and posed a strong challenge to Freud's lack of attention to the socio-historical context in configurations of gender. Developing Chodorow's [13] argument against Freud's theorizing of the Oedipal stage as determining gender, they argue that, from the very beginning of life, the child's relationship with his/her mother arises from a social context in which women are expected to be the main carers of children and others. This expectation is present in how the

---

[6] Mitchell, Juliet, *Psychoanalysis and Feminism: Freud, Reich, Laing, and Women*, (Harmondsworth: Pelican Books, 1975).

[7] Dinnerstein, D., *The Rocking of the Cradle and the Ruling of the World*, (London: Souvenir Press, 1978).

[8] Chodorow, N.J., *The Reproduction of Mothering*, (Berkeley and Los Angeles, CA: University of California Press, 1978).

[9] Eichenbaum, L. and Orbach, S., *Understanding Women*, (Harmondsworth: Penguin, 1985).

[10] Benjamin, J., *The Bonds of Love: Psychoanalysis, Feminism, and the Problem of Domination*, (London: Virago, 1990).

[11] Irigaray, Luce, *This Sex Which Is Not One*, translation by C. Porter, (Ithaca, NY: Cornell University Press, 1985).

[12] Kristeva, Julia, "Women's Time" in T. Moi (ed.), *The Kristeva Reader* (Oxford: Basil Blackwell, 1986), pp. 187 - 213.

[13] Chodorow, *ibid.*

mother touches, holds, feeds, and communicates with her girl baby. In theorizing women's ego development, Orbach and Eichenbaum claim that the ego boundaries between mothers and daughters, since they are of the same gender, are more blurred. This arises from their identification with one another and also from the social expectations of mothers to produce daughters whose priority will also be the care of others as mothers or carers. In their relationships, girls consequently experience themselves, consciously and unconsciously, as merged with and continuous with others. Furthermore, girls develop an unconscious internal boundary, splitting off those qualities which are regarded as unacceptable in women, such as strength, anger and independence in thinking and in action.

Eichenbaum and Orbach's conceptualization of the unconscious is a radical one, theorized as continuous with the wider social context, reflecting its social structure and its organization of gender. This conceptualization stands in strong contrast to notions of the unconscious as containing raw and primitive forces, which are assumed to be ahistorical and asocial in their origins. They are sensitive to the particular social pressures on girls and women at different stages of their lives, from infancy through childhood, adolescence and adulthood, and they include in their analysis the role of the media and cultural attitudes shaping girls' and young women's identities, consciously and unconsciously.

Eichenbaum's and Orbach's theorizing is, however, constrained by the over-valorization, in line with other Object Relations theorists, of the spatial metaphors of "inner" and "outer". In so doing, they assume a split between the subject and the world, the very split which their theorizing challenges. Self-understandings in terms of an "inner world" arise from a desire, particularly in the psychoanalytic relationship, for recognition of what is individual, personal and private in a person's experience. As Scarfone argues in relation to the notion of phantasies arising from within or from a "prehistoric past",[14] this split is problematic if the status of a psychoanalytic theory is uncritically assigned to a particular self-description. If an individual experiences what is individual to them as "inner", this does not mean that this dichotomy exists, or should exist, for everyone universally. As will be argued below, an analysis which, by contrast, emphasizes the temporality of the subject, allows for more specific interpretations of the shifting conscious and unconscious relationships that subjects "live" in the world.

Eichenbaum and Orbach conclude that, for a shift to occur in the binarism of gender roles, men and women must equally be involved in childcare, but they rely on a biologistic position which repeats the binarism. Their claim is that because mothers and daughters share the same biological sex, particular patterns of relating are reproduced. Yet this claim does not

[14] Scarfone, D., "Laplanche and Winnicott Meet ... and Survive", in L. Caldwell (ed.), *Sex and Sexuality: Winnicottian Perspectives* (London: Karnac, 2005), p. 37.

sufficiently take into account the individual circumstances of mothers in very diverse socio-historical contexts, class, and cultural backgrounds; rather, it raises the question as to whether, if the father is necessary for a developing girl's sense of autonomy, all daughters of lesbian parents are destined to particularly merged relations with others. Clearly, Eichenbaum and Orbach do not hold this view, given their openness to interpretations of lesbian identities as non-pathological and their critiques of how women's sexualities have been subjected to control in patriarchal cultures.[15] Indeed, Orbach and Eichenbaum's theorising of the presence of the father as a necessary condition for changes in the organisation of gender to occur throws their critiques open to question in their own terms.

Orbach and Eichenbaum's theorising is acutely relevant to the experiences of a great number of white middle-class women born in the late 1940's and 50's in Britain and the U.S., as evidenced by my psychoanalytic practice. This was a time when women were being stridently encouraged, through institutional and discursive practices such as education, medicine, and psychoanalysis, to relinquish the work and activities they had enjoyed in war-time and become full-time wives and mothers. In his novel *The Hours*, Cunningham describes this through the reflections of his character Laura:

> Because the war is over, the world has survived, and we are here, all of us, making homes, having and raising children, creating...a world of order and harmony where children are safe (if not happy), where men who have seen horrors beyond imagining, who have acted bravely and well, come home to lighted windows, to perfume, to plates and napkins.[16]

The daughters of Laura's generation of women suffered greatly from the notion that to become a woman was to become a mother or a carer like their own mothers. Furthermore, the mothering promoted as beneficial to families in post-war Britain and the US emphasized, as Cunningham conveys, detailed attention to physical care and comfort in the home, often at the expense of children's happiness and the development of their individuality.

Orbach and Eichenbaum's theorizing developed within a specific socio-historical context, namely the early 1980's, and they reflect on the lives of women whose childhoods extended back some decades before. It is heartening that since the time at which they were writing, and that owing to the influence of their work, there have been a great number of historical shifts

---

[15] See also Stacey and Biblarz's 2001 study of the effects of the sexual orientation of parents on children, which concluded that there were no differences between children raised by heterosexual parents and those raised by lesbian or gay couples. Stacey, J. and Biblarz, T. J., "(How) Does the Sexual Orientation of Parents Matter?" *American Sociological Review*. Vol. 66, (April 2001): 159 - 183.

[16] Cunningham, Michael, *The Hours* (London: The Fourth Estate, 1999), p.42.

which have profoundly affected women's lives. For example, adolescent girls' shame and embarrassment about their bodies and sexualities is less widespread and women are feeling more confident in defining their own sexualities. In addition, men are feeling more able to develop closer emotional ties with their children and to accept their own emotional vulnerabilities as well as their sexualities. The availability of AID and changes in adoption laws has had a significant effect on how lesbians experience their sexualities, both consciously and unconsciously. This change, of course, has had different meanings for every lesbian, depending on, for example, their age and their attitudes to parenting. Historical and cultural shifts such as these have radical effects on women's subjectivities, whether they constrain or open up different possibilities of existence.

Across the channel from the London Women's Therapy Centre founded by Eichenbaum, Orbach and other women, Kristeva, a Bulgarian born psychoanalyst living in Paris, explicitly addresses the question of time in relation to women's subjectivity. Kristeva's theorizing of language is influenced by Lacan's thinking. However, she extends his concepts of the "imaginary" and "symbolic" to include her notion of the "semiotic", a modality which is neither temporal nor spatial. The origins of the "semiotic" are in the pre-Oedipal, and according to Kristeva, is already indicated in Freud's concept of the primary process. The "semiotic" refers to an articulation of the pulsions or drives, predominantly oral and anal, and is discernible in the ruptures in the grammatical rules of language, through metonymy and metaphor. The drives are gathered in the "chora", "a non-expressive totality" which is both "mobile and extremely provisional."[17] The "chora" is organized through its orientation to the mother's body, and through familial and social constraints, not via the "law" which relies on symbolic language. Kristeva emphasizes that the "chora" is neither a sign nor a signifier; she equates it with vocal or kinetic rhythm whereby, for example music, according to her, is a non-verbal signifying system produced through the semiotic.

Kristeva's emphasis is on "abjection" and the separation from the pre-Oedipal mother, rather than castration. She describes abjection as the "violent, clumsy breaking away, with the constant risk of falling back under the sway of a power as securing as it is stifling";[18] yet also timeless, a time of "oblivion".[19] "Abjection" precedes the intervention of the father and the introduction of the child into symbolic language, through which the "chora" and the infant's connection to the maternal is repressed, albeit precariously.

---

[17] Kristeva, Julia, *Revolution in Poetic Language*, (New York and Guilford, Surrey: Columbia University Press, 1984), p.25.
[18] Kristeva, Julia, *Powers of Horror: An Essay On Abjection*, (New York: Columbia University, 1980), p.13.
[19] Kristeva, *ibid*, p. 9.

However, "abjection" inevitably returns in the interruptions, gaps, and rhythms within symbolic language.

Kristeva maintains Lacan's notion that it is the father who introduces the growing child into the symbolic order: the tie with the mother is severed and the child forced to accept its castration. In "Women's Time" she argues for an extension of the notion of castration (beyond the penis as main referent) to signify, for men and women, all that limits totality, namely "the break indispensable to the advent of the symbolic".[20] However, this argument does not lead Kristeva to substantially question her equation between the father, the symbolic order, and time itself: "There is no time without speech. Therefore there is no time without the father...[The father is]...sign and time".[21] The time of the father is linear time.

In "Women's Time", Kristeva seeks to redress the emphasis accorded to space in the theorizing of psychoanalytic theory, anti-psychiatry, matriarchal religions, and in Plato's philosophy. She differentiates between "monumental time"[22] which is "all-encompassing and infinite like imaginary space"[23] and associated with the maternal, and linear historical time. She argues that traditionally, female subjectivity has been connected with the former and also with the "cyclical time" of nature and biological rhythms.[24] She equates linear time with obsessionality and the "hysteric" (male or female) in flight from memory with cyclical and monumental time.

Kristeva claims that the struggles of suffragists and existential feminists for equal rights with men signified the desire for a place in linear time. Consequently, she argues, they rejected qualities such as the maternal which might jeopardize their inclusion in that time. She describes the next generation of feminists from 1968 on, by contrast, as repudiating linear time in favor of fluidity and plurality and, furthermore, a valorization of mythical memory and cyclical and monumental time. Through their writing and artistic productions (towards which Kristeva is scathing), these feminists were attempting to articulate bodily and intrasubjective experiences which had been previously silenced.

Kristeva regards the lesbian's or "homosexual" woman's position as characterizing a problematic within both linear and monumental temporalities. In *About Chinese Women* she presents the lesbian as one of the "supermen",[25] identifying with masculine values and, in "Women's Time", as destructively rejecting the "paternal function"[26] along with terrorists and single mothers, if she becomes a mother. Kristeva concludes with the offer of an alternative to

---

[20] Kristeva, "Women's Time", *op cit,* p. 198.
[21] Kristeva, Julia, *About Chinese Women*, (London: Marion Boyars, 1977), p.35.
[22] Kristeva, "Women's Time", *op cit,* p. 189.
[23] Kristeva, *ibid,* p.191.
[24] Kristeva, *ibid,* p.192.
[25] Kristeva, *About Chinese Women*, *op cit,* p. 37.
[26] Kristeva, "Women's Time", *op cit,* p. 205.

those she characterizes as belonging to the first two generations of feminists. She argues for a position which, she claims, allows for the "*singularity* of each person" and "the multiplicity of each person's possible identifications".[27] As an ethics, it involves drawing on the "cutting edge" of the symbolic to deconstruct the power-positions inherent within it and to open out possibilities for *jouissance*, responsibility, challenges, and differences.

Kristeva's explicit engagement with the themes of time, language, and gender appears to offer a promising analysis of the socio-historicity of the subject which might allow for differences (as opposed to universals). She is particularly original in her consideration of how different genders may be inserted into different times. Her account is, however, permeated with contradictions. Although Kristeva emphasizes the early relation with the maternal body, her perspective relies on the Oedipus complex as pivotal. She unquestioningly equates femininity with the maternal, and the symbolic with the father's intervention, while implying the possibility of a deconstruction of this equation. She assumes that women can insert themselves into a different time without questioning how the very equations she makes (between femininity and maternity; between fatherhood and the symbolic) paradoxically have the effect of maintaining that which she criticizes. Women who do take up particularly radical challenges to these positions (such as lesbian mothers or professional women who choose not to have children) are denigrated by Kristeva. This fact stands in strong contradiction to her argument for the possibility of a time which can allow for individual differences and a plurality of identifications which are free of the constraints of both male and female sexism.

The Time of the Political Psyche

Introducing the notion of "the father of whatever sex", Jungian psychoanalyst Samuels[28] goes further than Kristeva in opening out possibilities for multiple identifications in relation to gender and sexuality. His concern is not so much with the question of language but, rather, with addressing notions of "inner" and "outer" in the Anglo-American psychoanalytic literature. Challenging the rigidity of such distinctions in *The Political Psyche*, he argues that "subjectivity and intersubjectivity have political roots; they are not as 'internal' as they seem....the demarcation between the inner world of psychology and the outer world of politics has no permanent existence."[29] Although Samuels does not explicitly acknowledge it, it is striking that his analysis relies upon a recognition of human temporality. His work contains many references to time and to the socio-historical contexts of particular

[27] Kristeva, *ibid*, p. 210.
[28] Samuels, A., *The Political Psyche*, (London and New York: Routledge, 1993).
[29] Samuels, *ibid*, p. 4.

theories, including his own. The influence of Jung's recognition of the significance of the present and future, as well as the past temporal dimensions, is evident in the work of Samuels and allows him to develop a more complex theorizing of the subject's socio-historical and political specificity. He particularly challenges object relations theory, with its causal and sequential explanations in relation to origins, emphasizing the importance of the dimension of the future and potential for political change. Instead, Samuels argues, the question for the psychotherapist should be: "Where's it going? What does it mean? Not what are its roots?"[30] The notion of "development" is not natural; it is a fantasy and it is contingent. He argues that the father-infant relationship is not universal but subject to mutability within specific socio-historical contexts.

Samuels does, however, find it difficult to sustain his critique of the assumed division between inner and outer realities. On the last page of *The Political Psyche*, he refers to his struggle to "negotiate a tension between respect for historical specificity and context on the one hand and the limitless, playful vitality of imagination and fantasy on the other".[31] He states explicitly that psychic and political realities are different and implies, therefore, that freedom in imagination relies on its being ahistorical, acultural, and alinguistic. Samuels' conclusion risks replicating the very split that he challenges at many other points in the book. Nonetheless, Samuels' work raises challenging questions as to whether and how human imagination can exist outside or beyond history, language and culture; also, whether such a recognition of its historical and cultural specificity is necessarily restrictive?

Desires, Identities, and Historicity

A radical theorizing of the historical specificity of subjectivities has emerged predominantly in relation to the significance of the emergence of lesbian and gay identities. For example, the contemporary philosopher Butler argues that many philosophical accounts search for an "internal feature" of the subject which guarantees the continuity of their identity across time. She contends that the stability of the term "person" is severely challenged by the "cultural emergence of those 'incoherent' or 'discontinuous' gendered beings who appear to be persons but who fail to conform to the gendered norms of cultural intelligibility by which persons are defined".[32] Thus, "identity" is a normative concept which is itself socially and historically specific. Of particular relevance is the theorizing of analysts O'Connor (also a contemporary philosopher) and Ryan in *Wild Desires and Mistaken Identities, Lesbianism and Psychoanalysis*. Influenced by Merleau-Ponty and Foucault, O'Connor and Ryan challenge psychoanalytic notions of same-sex sexualities

---

[30] *ibid*, p. 74.
[31] *ibid*, p. 339.
[32] Butler, Judith, *Gender Trouble*, (London: Routledge, 1990), pp.16-17.

as necessarily pathological, arguing that such an interpretation arises from a reliance on "over-arching and universal metapsychological concepts", [33] which do not acknowledge socio-historical or socio-cultural specificity.

O'Connor and Ryan present a thorough and wide-ranging critical analysis of how lesbianism is positioned within psychoanalytic theorizing and practices. Acknowledging the particular socio-historical contexts which have produced this theorizing, their work vividly demonstrates how the issue of homosexuality within psychoanalysis has been "largely split off, inadequately discussed and understood, subject to rigid and sometimes attacking theorizing, and to excluding practices".[34] Central to their critique is a challenge to the notion of the Oedipus complex as universal. Drawing on Fanon, they argue that it does not allow for the diversity of patterns of child-rearing, historically and culturally, and also assumes a split between desire and identification: to be a (mature) man, one has to desire a woman.

O'Connor and Ryan are innovative in drawing on contemporary European philosophy with its emphasis on historicity and on the importance of language; in their view there is no "true" sexuality that lies outside language. They argue that the meanings of both homosexuality and heterosexuality are not "exhausted by reference just to intra- or inter-psychic events; they both denote significant aspects of the "deployment of sexuality" –to use Foucault's term– within the social order".[35] For Foucault, there is no body prior to history or to discourse; it is "produced" by and saturated with historical forces. Designations of the body and of sexuality as, for example, "natural" or "unnatural" are not factual distinctions given at the level of the body. These are interpretations which arise from the dominant discourses of the time, whether religious, medical, pedagogic, psychoanalytic etc. As O'Connor and Ryan argue in relation to psychoanalytic theories of sexuality, theories are never value-free. They arise from within particular socio-historical contexts and are saturated with the values of particular people at different times.

O'Connor and Ryan emphasize the multiplicity of sexualities and sexual identities and how these are produced from within particular socio-historical contexts. They argue that there are no essential features of lesbian desire which are necessarily shared by lesbians but acknowledge that homophobia may "constitute or contribute to [its] forms of expression".[36] An example of this, as I suggest elsewhere,[37] is the difference between the same-sex sexuality of upper middle-class white women in the 1920's, such as

[33] O'Connor and Ryan, *Wild Desires and Mistaken Identities, op cit*, p. 15.
[34] O'Connor and Ryan, *ibid*, p. 9.
[35] O'Connor and Ryan, *ibid*, p. 21.
[36] O'Connor and Ryan, *ibid*, p. 234.
[37] Ellis, Mary Lynne, "Who speaks? Who listens? Different voices, Different Sexualities" in J. Schaverien (ed.), *Gender: Countertransference and the Erotic Transference*, (Hove, E. Sussex: Routledge, 2006), pp. 123-142.

Virginia Woolf and Vita Sackville-West, and that of a black working-class young woman who defines herself as "queer" and enjoys "gender bending" contemporaneously.

Race in Time

Fanon's critique of the universality of the Oedipus complex, in his book *Black Skin, White Masks*, is supported by his evidence that in the French Antilles at that time "97 per cent of the families cannot produce one Oedipal neurosis",[38] and that every neurosis therefore has its source in the cultural context. Fanon is explicit from the beginning of the book that his analysis of black identities and the sources of racism includes both the socio-cultural (the context of place), and simultaneously, the socio-historical. In his warning to the reader, "I do not come with timeless truths",[39] he acknowledges the historical specificity of his own work.

Fanon's theorizing is influenced by French existential and phenomenological thought; he studied and worked in France as a psychiatrist before being assigned to a hospital in Algeria during the war of independence from the French. References to time in all its dimensions run throughout Fanon's conceptualization of subjectivity: the time of previous generations, the time of childhood, present time, and future time. His emphasis on temporality challenges the notion of a reductionistic "black psychology", one which he keenly repudiates. Fanon also argues against the temptation to revive or to attempt to romanticize the richness of previous, forgotten civilizations.

Nevertheless, according to Dalal, the fact that Fanon implies in his work the existence of a "pure 'race' of Negroes before the Diaspora"[40] signifies a weakness in his position. However, not withstanding this critique, it is important to emphasise Fanon's conviction that he will not make himself "the man of any past…I do not want to exalt the past at the expense of my present and my future".[41] The past has only relevance as a support for solidarity in the struggle against future subjugations of any group, and he warns against being determined by the present. He emphasizes the future dimension of subjectivity: "In the world through which I travel, I am endlessly creating myself" and "I am part of Being to the degree that I go beyond it".[42]

Fanon analyses the various culturally and historically specific influences on the development of black people's subjectivities, particularly in

---

[38] Fanon, Frantz, *Black Skin, White Masks*, (London: Pluto Press, 1986), p.152.
[39] Fanon, *ibid*, p.9.
[40] Dalal, Farhad, *Race, Colour and the Processes of Racialization*, (Hove, E. Sussex: Brunner-Routledge, 2002), p.95.
[41] Fanon, *ibid*, p. 226.
[42] Fanon, *ibid*, p. 229.

colonized cultures such as the Antilles where he grew up. He traces the role of magazines, comics and stories in the development of black children's identities. Produced mainly by white people (the colonizers), they depict the characters of savages, evil spirits, and the devil as black people or Indians. Fanon argues that the black (boy) child invests his own aggression in the white hero of the story and identifies with him. Since he does not see himself as black ("black people live in Africa"), but as an Antillean, it is only when he moves to Europe and faces overt racism that he realizes, in Fanon's words, "he is a Negro"[43]. Although Fanon is writing in the 1950's and addresses the particular issues of black identification for young black men, many aspects of Fanon's analysis continue to be relevant in Europe and the U.S. today. Contemporary children's literature contains fewer stereotypes, but there continues to be a dearth of black people in positions of power, including in the media. Education services are failing to address the needs of young black men and success is equated with being white. However, owing to the increasing strength of demands from black people for equality, together with an acknowledgement of differences, there is a much greater diversity of black identities, including those with high social status, available to black people in Europe and America contemporaneously.

Fanon analyses the unconscious anxieties on which anti-black racism hinges. Black people are associated with the biological, the instincts and genitality: "one is no longer aware of the Negro but only of a penis; the Negro is eclipsed. He is turned into a penis".[44] White men attribute their ideal of sexual potency to black men and seek revenge for their apparent possession of this genital superiority. In Fanon's theorizing of white women's fear of black men's sexuality, however, contradictions arise. His interpretation of the fear as primarily an expression of early developmental anxieties, contrasts strongly with his emphasis on the socio-historical and socio-cultural in his theorizing of black subjectivities. Drawing on the work of Helene Deutsch and Marie Bonaparte, Fanon claims that the little white girl (aged five to nine) observes her father beating a sibling rival. When he does not take up her unconscious aggression (previously directed clitorally towards her mother) she invests this in black men, in accordance with the prevailing discourses of the culture. However, its original source is the infantile need to "disembowel the mother", [45] which Fanon conveys as implicitly ahistorical and acultural. He concludes that the unconscious wish is that the black man "rip me open as I would have ripped a woman open".[46] The fear of being raped thus arises from the wish to be raped.

---

[43] Fanon, *ibid*, p.148.
[44] Fanon, *ibid*, p. 170.
[45] Fanon, *ibid*, p. 179,
[46] Fanon, *ibid*, p. 179.

Fanon does not offer any argument to support his retention of the Oedipus complex as central to his theorizing of white European women's sexual development, despite his fervent critique of its supposed universality. He uncritically accepts Freud's account of female development (expanded on by Deutsch and Bonaparte) through negotiation of her "double Oedipus complex"[47] and the attainment of "mature" vaginal sexuality. In relation to black women's "psychosexuality" he admits, "I know nothing about her,"[48] except that her fears are of the "all-but-whites" such as Senegalese men, whom she regards as inferior.

The contradiction between Fanon's interpretation of white women's racism as arising from early infantile drives, and his emphasis on the socio-historical and socio-cultural in his accounts of the shaping of black identities is striking. His position is particularly surprising given his critique of the psychoanalyst Manoni's assumption that a "dependency complex"[49] exists in colonized people which predates, and makes them susceptible to, colonization. However, his detailed analysis of the role of power and discourses in the production of socio-historical and culturally specific black identities was extremely radical for his time, anticipating Foucault's later analysis of the production of sexualities.

Fanon's analysis extends back to previous generations of black people, including their experiences of slavery. Fletchman-Smith[50] draws on her clinical experience with black men and women in analysis to support her argument that some of the dynamics of their intimate relationships contemporaneously can be traced back to the time of slavery and echo those which occurred between slaves and also between slaves and slave-owners. Fletchman-Smith does not claim that the transmission of these dynamics occurs in any biological or archetypal way. Rather, she implies, particular notions of intimacy are generated within and through parent-child relationships consciously and unconsciously from one generation to the next. Neither does she claim that there is one "black psychology". This claim would deny the diversity of the cultural backgrounds of people who define themselves as black, and assume that their conscious and unconscious experiences of their identities are fixed in time.

Aaron,[51] a Welsh literary theorist, shares Fletchman-Smith's interest in the cross-generational transmission of patterns of behavior. She describes how restrictions on Welsh women's sexuality inculcated by the colonizing

---

[47] Fanon, *ibid*, p. 178.
[48] Fanon, *ibid*, p. 180.
[49] Fanon, *ibid*, p. 85.
[50] Fletchman-Smith, B., Unpublished paper. Presented at Women's Therapy Conference, London, June 1992.
[51] Aaron, J., "Finding a Voice in Two Tongues: Gender and Colonization", in J. Aaron, Teresa Rees, Sandra Betts, and M. Vincentelli (eds.), *Our Sisters' Land: The Changing Identities of Women in Wales*, (Cardiff: University of Wales Press, Aaron, 1994).

culture (England) outlived the retention of these in English culture. She refers to a Welsh language text, which describes how a woman's guilt at unknowingly having had an affair with a married man was nearly as acute in 1966 as if it would have been in the late nineteenth century. During this latter period, Welsh non-conformity had introduced a strict policing of women's sexuality in order to prove to the English colonizers that Welsh women were not, as they believed, promiscuous and wild in their sexuality. This self-policing had continued to be passed down over generations and to have profound effects on women's sexual lives.

Ambiguity, Language and Embodiment

I shall now focus on some of the resonances between the texts of two writers from the fields of philosophy and literature, Merleau-Ponty, the twentieth-century French philosopher, and Eva Hoffman, the contemporary literary theorist and novelist. The themes of time and language are central in the work of both and their particularly extensive analysis of the socio-historical specificity of subjectivity is highly relevant to psychoanalytic practices.

In the *Phenomenology of Perception*,[52] Merleau-Ponty theorizes how we are embodied through and by the passage of time, including our anticipation of the future. He argues that subjectivity is always contextual and his theorizing is rooted in his challenge to Cartesian mind-body dualisms, including distinctions between "inner" and "outer", "internal" and "external". He criticizes the claims for universality in psychoanalytic theories with their failure to take into account socio-cultural and socio-historical contexts. Eva Hoffman's autobiographical work, *Lost in Translation*,[53] is a highly personal account of her reflections, questions and struggles as Polish and Jewish, as a girl and a young woman, through her shifts between different cultures across decades of radical political changes. The individuality and originality of her writing emerges from the intricate interweaving of the specific temporal and cultural contexts through which she moves. Some of Hoffman's reflections reveal her engagement with psychoanalysis (as both an academic and also, later, as a patient) but her work goes further than much psychoanalytic theorizing in addressing questions of time and subjectivity. Although there are significant differences in the views of Merleau-Ponty and Hoffman on the question of the continuity of the subject's identity through time, they both share an understanding of the subject as socio-historical.

Lost in Translation ends with the words: "Time pulses through my blood like a river. The language of this is sufficient. I am here now."[54] These words also echo Merleau-Ponty's conception of subjectivity as temporality

---

[52] Merleau-Ponty, Maurice, *The Phenomenology of Perception*, translation C. Smith, (Routledge and Kegan Paul, 1986).

[53] Hoffman, Eva, *Lost In Translation A Life in a New Language*, (London: Vintage, 1991).

[54] Hoffman, *ibid*, p. 280.

itself: "we must understand time as the subject and the subject as time".[55] Hoffman, "catapulted across so many generational divides, backward and forward at the same time",[56] is acutely sensitive to the temporal shifts in her experience between past and present time, and her ambitions for her future. Folded into her experience are the stories of the times of her parents' lives, their experiences as Jews before and after the First World War, their suffering, the enormity of their losses, their struggles to survive.

Throughout her autobiography, it is evident that Hoffman's most private reflections are deeply rooted within, and shaped by, the various socio-historical and, therefore, political contexts in which she lives and in which her parents lived. Writing of her childhood in post-war Poland, she recalls falling asleep as a small child amidst conversations which "weave in and out", such that she has the "sense that I am immersed in a stream, and that when I wake up there will be more stuff, more talk, more life".[57] For Hoffman, this context is not "added on" to personal experience or viewed by her simply as the "external" world. It forms the texture of her experience: "politics infiltrates daily lives in the most porous ways...we kids pick up reverberations of half-finished political phrases like bat signals".[58] These "reverberations" include the times of the stories of her parents' lives before her birth, and, most recently, the political atmosphere of terror, killing, and flight from the holocaust. As a small child, Hoffman is intent on searching the world for her mother's younger sister, who was killed in a gas chamber, in the hope of healing her mother's grief. She believes it is a "matter of honour"[59] to remember, yet she is too young to understand what she remembers. Moreover, to relive it all with her mother is a way for her to "atone for what happened".[60]

The stories of Hoffman's father eventually emerge out of the silence on which he has relied for so long for his sense of dignity. By then, his experiences, with more distance, have become "fables";[61] it is a friend of his who reveals to Hoffman (when she is an adult) the profundity of his suffering as he carried her mother, who had had a miscarriage, on his back, trudging several kilometers through snow. This friend's account weighs on her, forcing her to acknowledge "that this – the pain of this – is where I come from, and that it's useless to get away".[62]

Hoffman's account of how she absorbed the shifting political climate together with the stories of her parents' past (which were specific to

[55] Merleau-Ponty, *op cit*, p. 25.
[56] Hoffman, 1989, *op cit*, p. 250.
[57] *ibid*, p. 56.
[58] *ibid*, p. 57.
[59] *ibid*, p. 24.
[60] *ibid*, p. 24.
[61] *ibid*, p. 23.
[62] *ibid*, p. 25.

the particular socio-historical context in which they had lived), resonates with Merleau-Ponty's theorizing of "sedimentation".[63] He argues that, in our relations with others, we accumulate a background of shared meanings which inform our conversations without our needing to reassemble them on every occasion: they constitute a "sediment left by our mental processes, which enables us to rely on our concepts...without there being any need for us to resynthesize them."[64] This sediment is not, however, an inanimate layer within us, but a dynamic process, whereby every acquired meaning is revivified by a present thought.

This notion of "sedimentation" allows for the acquisition and retention of historical and cultural meanings which inform our ways of being in the world. Such meanings are, according to the theorizing of Jung, inherited in the form of the collective unconscious. For Merleau-Ponty these meanings are, instead, acquired through our intersubjective relationships through the languages of bodily gesture and verbal language (which may be transmitted from one generation to another through language). He argues that, in listening, we do not listen to a succession of words or facts, but rather follow the flow of a particular account. The story has a "style"[65] that is accessible to the subject because he/she is able to extend him/herself beyond his immediate experience to the events recounted. This account of listening emphasizes that listening is itself a form of embodiment through which "my world is expanded to the dimensions of that collective history which my private existence takes up and carries forward".[66] Merleau-Ponty argues furthermore, that it is impossible to delineate "where historical forces end and ours begin…since there is history only for a subject who lives through it, and a subject only in so far as he is historically situated".[67] In writing about class, he stresses that economic systems and societies are not "impersonal forces" and that the identity of, for example, being a proletarian, arises from how we carry these institutions within ourselves and experience living within their framework.[68]

Hoffman vividly describes how she becomes embodied through her experiences of alienation as a young immigrant in Vancouver, struggling to find a place in a radically different culture whose collective history she does not share. Out of time with this new culture and severed from her childhood in post-war Poland, she swirls in a world where her previous "grid of mental imagery",[69] the historical and cultural structuring of her perception which informed her previous existence, no longer has relevance. Now, in this

[63] Merleau-Ponty, *op cit*, p. 130.
[64] Merleau-Ponty, *ibid.*
[65] *ibid*, p.133
[66] *ibid*, p. 433.
[67] *ibid*, p. 173.
[68] *ibid*, p. 443.
[69] Hoffman, *op cit*, p. 134.

strange country, her vision is literally blurred. Her alienation is "inscribed in my flesh and face" and her shoulders are "bent with the strain of resentment and ingratiation".[70] An older Polish woman initiates her into the exacting requirements of early 1960's femininity in Canada: shaved underarms, plucked eyebrows, a bouffant, high-heels, lipstick and incessant smiles. Underneath this "elaborate packaging" her body is "stuff, sulky, wary".[71] When she meets up with a Polish woman friend, after living some time in America as a young woman, her embodiment as an American is accentuated for her. Her stride is more assertive than Danuta's, but her face is more controlled. Hoffman mourns the loss of its lively expressiveness which characterizes femininity in Poland. Her features and her friend's have been shaped by "different ideologies of femininity, different loci of restraint and expression".[72]

Hoffman rages at the impossibility of inhabiting Canadian English with the deftness and spontaneity she enjoys in her native Polish. The enormous cultural shift is simultaneously a temporal one: for Hoffman time stops as she desperately tries to freeze the past and resist the future: "I want to stop the flow. As a punishment, I exist in the stasis of a perpetual present, that other side of 'living in the present', which is not eternity but a prison. I can't throw a bridge between the present and the past, and therefore I can't make time move".[73] Later on, as a young woman in New York, Kennedy's assassination highlights how, for her, not knowing the historical background makes the "foreground" harder to grasp.[74] At college she finds that aspects of her history and her present, now she is older, do seem to converge with those of other young Americans, including her sense of "dislocation". But the historical sources of their dislocation differ from hers, and her American peers consequently refuse assimilation, while her own "uprootedness"[75] is underpinned by a longing for a sense of home.

Eva Hoffman recognizes that she has "been written in a variety of languages",[76] and it is an important moment when she dreams in English, discovering that words can be "a transparent medium in which I live and which lives in me".[77] Her realization echoes Merleau-Ponty's view that, in our listening, words become "the transparent envelope of a meaning within which (we) might live"[78] and not signs to be deciphered. Her descriptions challenge the dichotomies of "inner" and "outer" that permeate much of

---

[70] *ibid*, p.110.
[71] *ibid.*
[72] *ibid*, p.236.
[73] *ibid*, p. 117.
[74] *ibid*, p. 190.
[75] *ibid*, p. 197.
[76] *ibid*, p.275.
[77] *ibid*, p.243.
[78] *ibid*, p. 133.

psychoanalytic theorizing. Subjectivity is, for her, intrinsically socio-cultural and socio-historical. Later on in her book she muses that it is as though she has two unconsciouses, an American one and a Polish one. These both carry their own histories and the histories of others with whom she has been connected. This realization is a radical departure from notions of the unconscious as containing alinguistic or ahistorical impulses; for Hoffman the unconscious is historically and culturally constructed and this is evident throughout her descriptions of her life.

Hoffman's concept of the unconscious converges somewhat with that of Merleau-Ponty. He strongly criticizes naturalistic notions of the unconscious prevalent in psychoanalysis and, drawing on Freud's later work, presents a new formulation in which "conscious" and "unconscious" are not conceptualized as such discrete realms.[79] He questions how the unconscious can be made up of instinctual impulses if it selects what it admits to consciousness and makes us avoid thoughts or situations that we are resisting. For Freud, the super-ego is located in both the conscious and the unconscious. Merleau-Ponty concludes that the unconscious is "*not un-knowing* but rather an un-recognized and unformulated knowing that we do not want to assume".[80] It is this un-knowing, Merleau-Ponty claims, that Freud leans towards in his later work, and he suggests that it may be better formulated as "*ambiguous perception*".[81] A closer analysis of this concept reveals how it allows for more inclusive interpretations of the complexities of subjectivities in their conscious and unconscious aspects within particular socio-historical contexts.

Furthermore, Merleau-Ponty associates ambiguity with dialectic such that dualisms between, for example, inner and outer, mind and body, conscious and unconscious, are in an interdependent, rather than a contradictory or superseding relationship to one another. They may deny one another, but they rely on one another for their existence. Perception, for Merleau-Ponty, is always an act which gives meaning to data. As acts of ambiguous perception, dreams, humor, and mistakes are meaningful or condensed groupings of what he describes as "symbolic matrices, a language of self to self, systems of equivalences built up by the past".[82] They are embodied, linguistic and social, and therefore are shaped by specific socio-historical and cultural contexts. Their meanings are sometimes conscious and sometimes unconscious. As Merleau-Ponty argues, we know about them to

---

[79] Merleau-Ponty, Maurice, "Phenomenology and Psychoanalysis": preface to Hesnard's l'Oeuvre de Freud, translation A. L. Fisher, in K. Hoeller (ed.), *Merleau-Ponty and Psychology*, (NJ: Humanities Press International, 1994).
[80] Merleau-Ponty, Maurice, *Signs*, translation R. McCleary, (Evanston, IL: Northwestern University Press, 1964), p.229.
[81] *ibid.*
[82] Merleau-Ponty, "Phenomenology and Psychoanalysis", *op cit*, p.69.

the extent we do not know about them; and we do not know about them to the extent that we know about them.

Merleau-Ponty's critique of the notion of biologistic conceptions of the unconscious, and his notion of ambiguous perception opens up further possibilities. If there is a dialectical shifting between conscious and unconscious experiences, this must occur in time. The associative chains revealed through discussion of, for example, a dream may be more usefully viewed as connected horizontally back and forth across time, rather than vertically downwards into the so-called "depths" of "the unconscious" with its repressed impulses. Rather than *possessing* "an unconscious" in the form of a discreet chamber of instincts, it is our actions which possess conscious and unconscious meanings. As a result, terms such as "conscious" and "unconscious" may thus be better used adjectively or as adverbs.

Importantly, Hoffman's and Merleau-Ponty's accounts lead us in the direction of a new solution to the question posed by analysts concerned with the socio-historical and the socio-cultural; namely, how we can theories the relation between an individual's "internal" and "external" worlds. If, instead, we focus on the individual's temporality, on *how* they live, consciously and unconsciously, amongst the temporalities of others and how they carry forward the temporalities of those that have lived before them, the socio-historical specificity of their experiences is revealed as neither "inner" nor "outer".

Yet, it is important to emphasize that if, as is argued here, subjectivity is socio-historically specific, this does not mean that subjectivity is unalterably shaped, as if layer upon layer, by its socio-historically specific experiences. This fear is expressed by Hoffman when she furiously rails against the constraints of her socio-historical and cultural identities:

> How trite and tedious ... to see oneself as a creature formed by historic events and defined by sociological categories. I am a Jew, an immigrant, half-Pole, half-American...I suffer from certain syndromes because I was fed on stories of the war...I haven't escaped my past or my circumstances; they constrain me like a corset.[83]

This perspective on herself arises, in part, from what she assumes is the perception of some rich Bostonians she encounters as a student. Reading Nabokov's fiction, she longs for the freedom of his pure, aesthetic and individualistic world, "untrampled by history, or horrid intrusions of social circumstance".[84] However, in her question as to whether it is aristocratic privilege which provides access to this world, she paradoxically

[83] Hoffman, *op cit*, p.198.
[84] *ibid*, p.197.

acknowledges the socio-historical and cultural specificity of the possibility of such access and also assumes that it can be transcended. Hoffman concludes that "perhaps it's not possible to transcend our circumstances entirely after all".[85]

From Hoffman's dilemma as to whether we remain defined by our histories and our cultures or whether we can transcend them, a third possibility emerges. This possibility is not articulated explicitly by her, yet it is lived by her, from moment to moment, in its joys and its discomforts: through and between the shifts and contradictions in her identities new possibilities and new configurations arise from complex matrices of identifications which stretch back and forth in time as well as across places. In the vulnerability of being an "outsider", she fights fiercely against the possibility of subjection to the dominant (American) culture. She moves between the voices of her Polish past, including her parents' creativity in their determination to escape from the Holocaust, her adolescence in Canada, and her new life, beginning as a student, in America. It is the interweaving of these different and sometimes conflictual voices across time, as well as place, which produces a critical questioning and forms the source of Hoffman's individuality, and originality - not impediments to it.

Indeed, the post-Colonial theorist Edward Said recognizes the creativity that can emerge from exile: "Exiles cross borders, break barriers of thought and experience".[86] Movement from one culture to another does not involve only a spatial shift; it is also temporal. He emphasizes, as Hoffman's work so vividly conveys, that every experience in the new culture "occur(s) against" the memories of those in the previous culture.[87] What is recalled of the past and how it is remembered also shape the individual's perception of the future. This conceptualization does not, as Said and Foucault emphasize,[88] assume that identities are continuous through time. Hoffman is, for example, explicit as to how the journey from Poland to Canada marks the end of "the narrative" of her childhood.[89]

Foucault, Said, and Hoffman all highlight the importance of gaps, discontinuities, and contradictions in allowing potential for new thoughts and the emergence of resistances to a dominant order. Their theorizing offers a solution to the question posed by Samuels as to how to "negotiate a tension between respect for historical specificity and context on the one hand and the limitless, playful vitality of imagination and fantasy on the other".[90] Eva

---

[85] *ibid,* p.198.

[86] Said, Edward, *Reflections on Exile and Other Essays*, (Cambridge, MA: Harvard University Press, 2000), p.185.

[87] Said, Edward, *Freud and the non-European*, (London: Verso, 2003), p.8.

[88] Foucault, Michael, *The Archeology of Knowledge*, translation A. A. Sheridan Smith, (London: Routledge, 1995).

[89] Hoffman, *op cit,* p.95.

[90] Samuels, *op cit*, p. 339.

Hoffman's autobiographical work implicitly cuts through this split, revealing it as being, itself, a cultural-historical construct.

Conclusion

In the analyst's awareness of the socio-historical as intrinsic to our conscious and unconscious experience there is, implicit in this, an orientation to the possibility of change: every moment produces a new context for a different possibility of experience. In the analytic dyad an acknowledgement of the socio-historical aspects of temporality can contribute to a richer and more complex understanding of an individual's experiences. Such interpretations should not be regarded as a "collapse into sociology", or assumed to be merely generalizing. As Hoffman's autobiography so vividly conveys, a recognition of socio-historicity, in contrast, individualizes the experiences that are brought to the psychoanalytic relationship.

I have highlighted how the familial constellations in which individuals grow up are shaped by specific socio-historical contexts and cannot be adequately theorized in terms of Oedipal (or pre-Oedipal) dynamics. Reductive interpretations of this type are generalizing and do not address the uniqueness of the positions within which individuals recurrently discover themselves, amongst matrices of criss-crossing identifications stretching back to the past and forward into the future.

Psychoanalytic attunement to the socio-historical requires an acknowledgement of difference and of change. This orientation in the analyst's relationship with the patient is vital for the emergence of new possibilities of being in the individual's life. In contrast to notions of "instincts" as raw, alinguistic impulses, or of the universality of the "archetypes" of Jung's "collective unconscious", this interpretation allows for multiple meanings of experience which shift as they are perceived from within different temporal contexts.

Through consideration of the temporality of the subject, and through abandoning the binary oppositions of the "external" as opposed to the "internal", a notion of subjectivity emerges in which the psychic and the social are integral to one another.

As I have argued through my analysis of the work of Merleau-Ponty, Foucault, Fanon, Said, and Eva Hoffman, psychoanalytic theory and practices require a new conceptualization of subjectivity as located in and moving between constantly shifting matrices of complex webs of connections, in which there are numerous possibilities of play, of creating and breaking connections, and of originality produced from the extraordinary crisscrossing of the fibers and intricacies of all of our relationships. Integral to such a conception is the recognition of the subject as temporal.

Intermittences: Merleau-Ponty and Proust on Time and Grief [1]

Patricia Locke

Proust's writing mimics time, with its flow, its looping back upon itself, its seeming eddies and sudden rush towards the future. Locke carefully considers in turn, affective time, as it is distinct from clock time, rational time, and historical time. "Intermittences of the Heart," a section spatially set off from the continuous flow of Proust's novel *In Search of Lost Time*, displays the narrator's realization that he is temporal, spatial, sexual, and that these interwoven dimensions of his existence are constitutive of human being. This awareness is triggered by disruptions in the field of being, like death, grief and violence. Affective time, for instance, presents itself as continuous, but not as a smooth sequence for the one who is caught up in grief. There is a suspension of the present, a folding of past into the present in ways that do not fit, a discontinuity of selves who are also grieved along with the one lost. Proust intended "Intermittences of the Heart" to be part of a three volume work modeled on three-fold experienced, lost and regained time. Locke examines how affective life "takes its own sweet time" with the help of Merleau-Ponty's phenomenology, especially as he spells out the interweaving of sexuality, space and time in Phenomenology of Perception.

For to the disturbances of memory are linked the intermittences of the heart.[2]

Time feels like a continuous flow pouring towards us, Merleau-Ponty claims in *Phenomenology of Perception.* The future broods like a storm on the horizon of my current life, and invisibly thickens the air with anticipation.[3] As an embodied being, I move towards the future, which already belongs to me. In fact, lived time arises from the relation between *me* and things that appear to change. I constitute time, and sense events moving "behind" me into a past that is still on my horizon. These inevitable spatial metaphors show that bodily movement initiates the coming into being of space and time. Yet in the chapter marked "Temporality", Merleau-Ponty adopts Husserl's schema that denotes time as "not a line, but a network of intentionalities."[4]

[1] This essay is dedicated to the memory of my uncle, Clayton A. Shupe, 1934-2007.

[2] Proust, Marcel, *In Search of Lost Time*, general editor Christopher Prendergast, especially *Sodom and Gomorrah*, translation by John Sturrock, (NY: Viking, 2004), p. 155. *À la recherche du temps perdu,* general editor Jean-Yves Tadié. (Paris: Gallimard, 1987), F III: 153.

[3] Merleau-Ponty, Maurice, *Phenomenology of Perception*, translation by Colin Smith, (NY: Routledge, 1962), p. 411. *Phénoménologie de la perception,* (Paris: Gallimard, 1945), F. 471.

[4] Merleau-Ponty, *ibid*, p. 417; F. 477. This temporal schema is borrowed from Husserl, *Zeitbewusstsein,* p. 21, who in turn borrowed the image from Brentano, Franz, *Sensory and*

This network allows me to experience the world as continuous, and events as carriers of meaning for me. The world fans out in various dimensions, and is receptive to my intentions.

This reciprocal interaction between self and world is centered in my body, which is fundamentally sexual. Merleau-Ponty stresses this point; the way in which I am a moving human being, able to take "root" in different settings, is my sexuality as a general power of being in relationships with others. He mentions this with reference to Freud's notion of "over-determination", by which actions have multiple meanings. [5] The heart, apparently, has reasons that reason may not fully comprehend. Sexuality as used here is bodily but not simply so, since I can be both sexual object and subject in the world. Not only expressed in overt ways, sexuality is laced throughout our intentions and activities as a whole.

Yet what of the intermittences of the heart that Marcel Proust describes, which stem from events that shock us by their unintelligibility? Disruptions in the field of being that don't make sense, such as death, grief, or violence, show themselves as punctual. One must use the past definite to describe these significant events, though their consequences cast long shadows. These moments seem to stop time's flow in a jarring way, to fix it as a butterfly pinned to a board. Nothing happens afterwards, since that which made "sense" or gave meaning to things no longer coheres from the perspective of the heart. Even if the world continues apace for others in the surrounding field, for the grieving person the event cannot be brought into relation with memories that made life have a narrative shape and flow.

Merleau-Ponty is interested in "true" time, just as the "true body' is the one we live, in contrast to clock time or the measured "objective" body moving through Cartesian space.[6] This true time is that which is woven with the heart beat and bodily gestures, the memories and expectations of particular individuals. Each person affects and is affected by time passing with a different tempo, within the compatible phenomenal structures that make us one human community. Merleau-Ponty states:

> We can get through *(pénétrer)* to the individual only by the hybrid procedure (*le procéde bâtard)* of finding an *example,* that is, by

---

Noetic Consciousness: Psychology from an Empirical Standpoint, Vol. III (London: Routledge & Kegan Paul, 1981), p. 37.

[5] Merleau-Ponty, *ibid,* p. 158; F. 185.

[6] "Constituted time, the series of possible relations in terms of before and after, is not time itself, but the ultimate recording of time, the result of its *passage*, which objective thinking always presupposes yet never manages to fasten on to. It is spatial, since its moments co-exist spread out before thought; it is a present, because consciousness is contemporary with all times. It is a setting distinct from me and unchanging, in which nothing either elapses or happens. There must be another true time, in which I learn the nature of flux and transience itself." Merleau-Ponty, *ibid,* p. 415; F. 475.

> stripping it of its facticity. Thus it is questionable whether thought can ever quite cease to be inductive, and whether it can assimilate any experience to the point of taking up and appropriating *(posséder)* its whole texture.[7]

We will approach time through the example of *À la recherche du temps perdu*, since Proust's writing mimics affective time, with its flow, its looping back upon itself, its seeming eddies, and sudden jerking forward towards the future. Proust's task is to "regain time." This lost time is tied to the lost ability to feel the feelings of youth while, as an adult narrator, to give them meaning. My question is how can true time, taken as continuous, tolerate felt discontinuities, the lost times?

In this paper, I will consider an episode of involuntary memory, a seeming temporal discontinuity. "Intermittences of the Heart", a section spatially set off within Proust's novel, displays the narrator's first realization that he is temporal, spatial, and sexual, and that these interwoven dimensions of his existence are constitutive of a human being. Proust had intended "Intermittences of the Heart" to be part of a three volume work modeled on three-fold experienced, lost and regained time. World War I disrupted the publication of *In Search of Lost Time*, which then expanded exponentially. "Intermittences" was thrust forward into volume four, *Sodom and Gomorrah*. Thus, the return of childhood affect in the context of adult sexuality is heightened, and the juxtaposition is based on violence and death. Marcel is a young man, and has already had childhood traumas, experienced dislocations in space and agonies of waiting for events that never occur. Nonetheless, the category of intermittences brings him into being as fully temporal, because through death he becomes fully relational with others *as* other.

Proust allows us to follow a thread of the narrator's life with his grandmother through the entire seven volumes, letting us see his youngest self that doesn't distinguish itself from another, or the present from the future; the self that sees the grandmother as separate, old and repulsive, and at last the self that considers the grandmother as finally dead and his own self with her.[8] Marcel feels genuine sorrow and only recognizes the significance of his

---

[7] Merleau-Ponty, *ibid*, p. 63; F, 76.

[8] See Proust, Marcel, *Guermantes Way*, translation by Mark Trehane, (NY: Viking Penguin, 2004), pp. 134-35; F, II: 439. Here is the description of Marcel seeing his grandmother as a distinct person for the first time, as if his eyes "function mechanically like photographic film." Marcel describes himself like a sick man who hasn't looked in the mirror for a long time, then "recoils" when he sees his "monumental red nose" reflected. He defines himself, "I for whom my grandmother was still myself—I who had only ever seen her with my soul, always at the same point of the past, through the transparency of contiguous and overlapping memories—suddenly, in our drawing room, which had now become part of a new world, the world of Time," sees age upon her. But this is still a mechanical view, not felt in the body. The overlapping memories help unite Marcel's snapshots of his grandmother: on the first trip to Balbec, with the meeting of Albertine (vol. 2), at home in Paris, ill after her stroke; her attempt to look healthy in

grandmother's death a year after her funeral. He thought he had loved her and grieved her, but when he revisits a room where he had been aware of her deep love, its absence strikes him down. His first visit to this beachside hotel room had been on the horizon of a sexual awakening; his return several years later confirms that his future will be bereft of unconditional love.

We are lulled by the meticulous beginning of the section, Marcel's amused absorption in word play and idle thoughts about the elevator boy, and then Proust shocks us with a short, sharp sentence fragment:

> "*Bouleversement de toute ma personne*; a convulsion of my entire being." [9]

This fragment fails to act. The reader feels a sympathetic uprooting, based on the shattered expectation of another luxurious Proustian sentence, full of details that would help us experience the passage of time itself. This jarring clause prefaces Marcel's first involuntary memory. As he wearily bends to undo the buttons on his boots after a long journey, he is convulsively thrown into the past time when his grandmother had knelt to help him. Shaken by sobs, and "filled with an unknown, divine presence," Marcel glimpses the disappointed face of his "true grandmother, the living reality of whom, for the first time since…she had suffered her stroke, I had rediscovered." [10] Marcel feels himself as desperately *needing* to be held in her arms, encircled by her protective love.

Proust's narrator pauses in this critical moment to note the "anachronism which so often prevents the calendar of facts from coinciding with that of our feelings." Marcel now realizes that he "was lonely and distressed, as he was long ago, but this time he no longer has someone ("for it was both me and more than me") to "restore him to himself."[11] The narrator reflects:

> "Now since the self that I had suddenly rebecome had not existed since that far-off evening when my grandmother had undressed me on my arrival in Balbec, it was, quite naturally, not after the day we were living, of which that self knew nothing, but—as if there were, in time, different and parallel series—without any break in continuity, immediately after that first evening in the past, that I adhered to the moment when my grandmother had leaned toward me. The self that I was then and which had vanished all that time ago, was once again so close to me that I seemed to hear still the

a photograph for him; her death, and the aftermath of increasing social preoccupations to crowd out these memories.

[9] Proust, *Sodom and Gomorrah*, p. 154; F, III: 152.

[10] Proust, *ibid*, p. 155; F, III: 153.

[11] Proust, *ibid*, p. 154; F, III: 152.

> words that had come immediately before, yet were no more than a dream…"[12]

This reanimated self lives in a series parallel to the one we had been reading about just a page before. A need was reborn, right when the possibility of fulfilling it was lost forever with his grandmother. This yields a contradiction: a return of need and tenderness *in him* "such as he had known them", and the certainty of nothingness. Thus, the affect from long ago enters the now as immediate, and is woven into the series that once was parallel.

Affective time presents itself as continuous, but not as a smooth sequence for the one who is caught up in grief. There is a felt suspension of the present, a folding of parallel pasts into the present in ways that do not fit, a discontinuity of selves who are also grieved along with the one lost. Reflection on the involuntary event loosens us from the thought of time as a causal chain. The event, whether joyful or painful, enables us to remember many other suppressed things—like Combray for the child Marcel, or the necessity of letting go of childhood to become an adult. Time and significance are one thing, says Merleau-Ponty, and are constituted by the subject him or her self. Even though an event "A" comes "after" event "B", they are both part of a web of relations that overlap in non-sequential ways. Something from the deep past can still be very meaningful now and color how we interpret events/objects/people in the present. And telling oneself—"that's over and done with''—may be untruthful and paralyze the natural mobility of temporal relations.

The words past, present, and future take their referents based on the position of the subject relative to his or her own life and placement in the world. The reader too feels a dislocation in temporal experience, here in volume four, *Sodom and Gomorrah.* The famous madeleine episode is related early in volume one, *À côte de chez Swann (Swann's Way)*, but it is chronologically later in the middle-aged narrator's life. Thus, the vivid madeleine episode, and its wake, affects our reading of the intermittences section. The madeleine experience was similarly preceded by a disheartened and weary attitude on Marcel's part, but the tea and cookie was offered by his mother, who by that time had adopted the habits and clothing of the deceased "good" grand-mother. In other words, she had changed from the unavailable mother of Marcel's memories of early childhood to become the solicitous grandmother. On first reading, we focus on Marcel's efforts to retain and reflect on the joyful involuntary memory triggered by the madeleine, and we pay little attention to the mother. The writer did not signal that this detail is important, and we are vague about the adult Marcel's living arrangements with his mother.

---

[12] Proust, *ibid*, p. 156; F, III: 154.

When Marcel reopens time in the boot button event, his experience is one of a vital present moment, which calls the recent past into question. For a moment, he hopes to restore not only the true grandmother, or his childhood entire, but himself. Merleau-Ponty observes that lived time "...exists only when subjectivity is there to disrupt the plenitude of being in itself, to adumbrate a perspective, and introduce non-being into it. A past and a future spring forth when I reach out towards them."[13] This subjective being is not a Kantian transcendental "I" that looks out at a constituted, spatially arrayed temporal progression, nor is the empirical body simply along for the ride. The world may be thought to be temporally continuous, but without a subjective perspective, it is not distinguishable from a perfectly still plenum. In a way similar to the spatial orientation the world receives from me as I move through space, gathering aspects to make a whole, the world in its temporal dimension is articulated by Marcel, trying to put together "true" unconditional maternal love with the seemingly ill-fitting gestures he receives from others. One recalls the crucial night of his childhood, when his mother slept in his room for the last time, trying to comfort the agitated child. She read aloud from George Sand's novel, *François le champi (François the Orphan)*, a present from Marcel's grandmother. However, she skipped all of the suggestive passages in this story of how an orphan grew in pure love for his adoptive mother and when grown was able to marry her. Thus, Marcel's sense of the narrative arc was confused at best. Sexuality and the phantasy of living forever with mamma is edited out of his childhood bedroom, even as his nervous temperament allowed him to remain childish and pampered by women with disappointment written all over their faces.[14]

Marcel shows us someone who posits himself *as* temporal, as human, when faced with the emotional reality of his grandmother's death. This self-positing takes place in response to events given to him by the world as an invitation to take up a situation within it. Only by taking up the task does a subject become him or her self, as a self-constituting maker of meaning. Marcel has benefited from his invalidism, evading moral responsibility for his toying with girls, and dwelling in an imaginary world where he hasn't yet fulfilled his promise as a writer. He has established a habit of kissing away the pain registered on the faces of those he has hurt for the express purpose of being able to kiss them. His grandmother, mother, and later his girlfriend Albertine prove their love by staying with him, despite his felt unworthiness to be loved. Once his grandmother is truly dead, the only one he continues to

---

[13] Merleau-Ponty, *op cit*, p. 421; F 481.

[14] After his insertion into time, Marcel goes so far as to reflect that "By dint of believing ourselves to be ill, we become so, we lose weight, we no longer have the strength to get up, we suffer from nervous enteritis. By dint of having tender thoughts about men, we become a woman, and a false skirt impedes our steps. In such cases, the *idée fixe* can modify (as in other cases our health) our gender." Proust, *Sodom and Gomorrah*, *op cit*, p. 300; F, III: 300-301.

wound is himself, with no opportunity to heal through kisses. He is at an impasse. Marcel must come to live in the present, able to directly encounter others.

Sexuality is present as an atmosphere that saturates his and our existence, and gives a layer of meaning to our every action. Merleau-Ponty devotes a chapter in *Phenomenology of Perception* to the non-contingent character of sexuality, and our reaching a life in relation.[15] He distinguishes, for example, a willful refusal to speak, and the inability to speak of a young girl whose parents cut off her romantic relationship. Her malady is below the threshold of the will, rooted in her body, and cannot be cured by reason or medication. Up to this point in life, Marcel had considered other people as extensions of himself, and just now enters the world of time. The position of Marcel's body (in the same hotel room, making the same gesture) for involuntary memory, and his reflection on it, incorporate temporality yet deny sexual components. This tension is the horizon that hovers around *À la recherche du temps perdu,* for even though the narrator is seeking time past, he is holding himself aloof from the time future that continues to advance towards him.

Just before the intermittences section, Marcel is stuck, as if looking at a flat screen projection of his past life with his beloved grandmother, and he cannot construct a fuller sense of the space rendered by her loss. He cannot live in a full present, with complex memories that change over time, illuminating this or that angle of who grand-mère had been. Merleau-Ponty reminds us that Marcel isn't simply caught in a binary opposition between absence/presence, or between forgetfulness/memory. Merleau-Ponty notes that a subject in psychoanalysis,

> ...knows what he does not want to face, otherwise he would not be able to avoid it so successfully. We do not understand the absence or death of a friend until the time comes when we expect a reply from him and when we realize that we shall never again receive one; so at first we avoid asking in order not to have to notice this silence; we turn aside from those areas of our life in which we might meet this nothingness, but this very fact necessitates that we intuit them.[16]

For over a year, Marcel has managed to avoid the challenge of his grandmother's absence and his self-without-her. Yet the superficiality of his social existence acts as a warding off, shaped in the exact contours of the rejected event. Merleau-Ponty points out that, "Proust can recognize the death of his grandmother, yet without losing her, as long as he can keep her

[15] Merleau-Ponty, *op cit*, p. 160; F, 186.
[16] Merleau-Ponty, *ibid*, p. 80-81; F, 96.

on the horizon of his life."[17] This "ambivalent presence" is a subconscious refusal of loss, like a man who acts as if he still has a leg that had been amputated during wartime. The missing response of someone who "ought" to be there supports the refusal. Marcel feels his grandmother as a part of himself and *cannot afford* to lose her. Yet without doing so, he will never be able to wholly put on his own boots.

The narrator links the disturbances of memory to the intermittences of the heart, as the sudden return of times lost cause affect to shift and to twist, just as a break in habitual behavior by a new feeling is often connected with a new meaning for past events.[18] The effort to hold still and wait is futile, especially for an undisciplined and self-involved young man such as Marcel. Desires clash, loyalties compete, and it is only in extreme situations that someone can persist along one line without external response. Our memories are pushed aside by selfish ordinary thoughts, but may reside in an "unknown region," akin to the Freudian unconscious. Paradoxically, while we live in the present moment, we are not fully alive to ourselves, and cannot know ourselves as wholes. Both his mother and grandmother tried to protect Marcel from knowing things about himself and from the world's dangers. He, in turn, tried to soften his grandmother's death, in part by denial, and in part by letting his mother assume this role.

All our gestures are charged with our sexual, temporal and spatial embrace of the world, or our own withholding of our intention towards it. When we refuse the future in grief, we close ourselves off from time, making knots and lacunae in its web. Yet those discontinuities are a part of the design. In the aftermath of the profound recognition of the finality of death, Marcel suddenly can emphasize with his mother. He has lost one of his own selves—the one cared for by grandmother—but now realizes that his mother has lost her *own* mother.

> Then, for the first time, and because I felt a sorrow that was nothing in comparison with her own, but which had opened my eyes, I realized with terror what she might be suffering. I

---

[17] Merleau-Ponty, *ibid*, p. 81; F, 96.

[18] "With a woman who does not love us, as with someone who has died, the knowledge that there is nothing left to hope for does not prevent us from going on waiting. One lives in a state of alertness, eyes and ears open; a mother whose son has gone on a dangerous sea voyage always has the feeling, even when she has long known for certain that he has perished, that he is just about to come through the door, saved by a miracle, unscathed. This waiting, depending on the strength of her memory and her bodily resistance, may enable her to last out the years that will eventually bring her to an acceptance of the death of her son, so that she gradually forgets and goes on living—or it may kill her." In referring to overlapping wishes that conflict with one another, Marcel continues: "...few of such wishes are absolute, at least not continuously so, one of the laws of human makeup being intermittence, which is further affected by the unpredictable recurrence of different memories." Marcel Proust, *In the Shadow of Young Girls in Flower,* translated by James Grieve, (NY: Viking Penguin, 2004), p. 166.

> understood for the first time that the fixed, tearless gaze…that she had had since my grandmother's death had been dwelling on this incomprehensible contradiction between memory and nothingness.[19]

The body acts as an expressive space: Marcel's mother deals with the loss by wearing her mother's clothes, reading her mother's books, staying in the same room in the resort hotel, in parallel to the gestures made by the deceased, who thereby is in some sense still alive in her. Marcel tries to keep her company, even though he realizes that his sense of the nothingness following a profound loss has been mitigated by his self-involved habits. Yet he feels terror, for he can truly feel for the first time what another person might be suffering. From now on, when he withdraws from taking responsibility for his actions, or intends to hurt or mislead others, it can no longer be said that he "didn't know" that he is enmeshed in an intersubjective world where his acts have meanings. The time of grief consequently turns out to be like planetary retrograde motion, where Venus, for example, appears to remain stationary or even move backwards, when inexorably there is forward motion seen from another perspective. The person who grieves cannot but enter this suspended or backwards looking pattern, but unlike a planet, whether he or she will take up the task of living forward again is an open question. Merleau-Ponty brings us to the point:

> It is of the essence of time to be not only actual time, or time which flows, but also time which is aware of itself, for the explosion or dehiscence of the present towards a future is the archetype of the *relationship of self to self*, and it traces out an interiority or ipseity. Here a light bursts forth, for here we are no longer concerned with a being which reposes within itself, but with a being the whole essence of which, like that of light, is to *make visible*.[20]

In *Sodom and Gomorrah,* Proust's phenomenological description of the time of grief makes visible the inescapable relation of self to self, whether conceived as regained or held in doubt. Marcel has shown us the first motions of self-aware time, a motion that turns towards others. Even when it appears that time has stopped and nothing more can happen, the self that remembers is invited to initiate genuine relations with others as other. He has come to see that his grandmother, assumed to be inside himself, had a life beyond him, and that now she has passed away. Insofar as the grandmother survives, it is in the memories within Marcel and his mother, acting towards a

[19] Proust, *Sodom and Gomorrah*, p. 167; F, III: 165.
[20] Merleau-Ponty, *op cit*, p. 426; F, 487.

future which includes her, whether by gesture or in thought. They make visible her past effect upon themselves, make choices with her in mind, yet now must bear up under her unresponsiveness.

Intermittences II: Albertine

The affect of grief wears off, and Marcel resumes his frivolous social life in the beach resort, now accompanied by Albertine, one of the little gang of girls he had met on his first trip to Balbec. Marcel struggles with jealousy, which in this case has a wide field of operation. He suspects that Albertine may be a lesbian, and thus doubts her every affectionate gesture towards a friend and every outing. He even holds her aloof from his male friends, just in case. He attempts to buy her attentions with presents, and deliberately manipulates her with improbable tales of his other liaisons, banking on her sympathy in his fictional love troubles. He wields the power of the invalid, whose delicate condition requires her attendance above all other obligations or desires. Yet this life and her place in it yield ennui at bottom, and time hangs heavily. He resolves to break up with her, to establish life anew, and then she casually mentions that she is an acquaintance of Mlle. Vinteuil and her "girlfriend, older than me, who was like both a mother and sister to me."[21]

This causes the sudden shock that Marcel identifies as a second intermittence, which alters the course of his life. The next two volumes of the novel will play out the consequences of her remark, for Marcel feels he has confirmation of her sexual tendencies and a resultant urgent need to hold her so close to him that she may be termed a prisoner. Like the love of a dead woman, a lesbian's romantic desire is impossible to attain. Though Albertine continues to show nothing but attentiveness towards Marcel, he recalls a long forgotten memory she cannot combat. He remembers resting as a young boy in the grass during a long walk near Montjouvain, the home of the composer M.Vinteuil and his wayward teenage daughter. Accidentally, he found himself watching from behind a bush, as MlleVinteuil and her girlfriend enacted a sadistic scene. Marcel listened in fascination near the open window, as the girls cruelly profaned a photograph of the recently deceased M. Vinteuil with their flirtatious display. He later considered that scene as a melodrama, played out by a tender-hearted girl who wanted relief from her deep mourning through evil, "a state so rare, so extraordinary, so disorienting, and to which it was so restful to emigrate."[22] This scene linked sensual pleasure with the "inhuman" or diabolical not only for Mlle. Vinteuil, but for Marcel. At this early stage of his life, his sexual desires are inchoate and mark out his solitude. As a young man, his desires bear the stamp of voyeurism and passive speculation about others' activities, without

[21] Proust, *Sodom and Gomorrah*, p. 499; F, III: 499.
[22] Proust, *Swann's Way*, p. 169; F, I: 163.

acknowledgement of his true desires and fear that he is not worthy of love. He interprets the return of the Montjouvain scene as his "torment—as my punishment perhaps, who knows, for having allowed my grandmother to die; suddenly rising up out of the depths of that darkness where it had seemed to lie forever entombed and striking like an Avenger, in order to inaugurate for me a new life," a life of continual suffering, a life weighed down by knowledge.[23]

Now, having to face the certainty that Albertine could not love him, he insists that she come to him in his hotel room to console him in his unexplained grief. She agrees to come, though annoyed at conforming to a seeming whim. This night his mother is asleep behind the thin wall that he used to knock on to let his grandmother know he needed her, with her comforting reply. Their rendezvous must be quiet, giving Marcel that undercurrent of the danger of being caught that also charged his earlier voyeurism. The hotel room that has seen so many interior changes in Marcel, while steadfastly containing the same furniture and sounds of the sea, witnesses him kiss Albertine's "powerful, golden, coarse-grained," neck as "innocently as if I had been kissing my mother, to still a childish unhappiness that I thought I would never be able to tear out from my heart."[24] This is the kiss that he wished to give his mother, long ago, to calm his anguish, when she was too busy with company to come upstairs to tuck him into bed. These juxtapositions of events from different temporal sequences force the narrator to wonder whether the sources of his deep unhappiness might be parallel to our responses to novelistic techniques. If he could close the book, or penetrate the page like jumping through a circus hoop, he might find release from grief. We the readers can close the book, but we simply exchange one train of experiences for another. Our dilemmas will not be easily resolved, since we are embodied and cannot take up permanent residence in imaginary or past worlds.

After Albertine leaves this night, Marcel cries softly, but is heard next door. The sun breaks over the horizon, signaling time's passage, as the door opens and his grandmother walks into the room. Or is it a dream? It is his mother, wearing the grandmother's dressing gown, sensitive to the morning and to his suffering. But Marcel cannot appreciate the dawn over the ocean, for it functions as a thin veil between him and the still vivid Montjouvain scene. The past is more real for Marcel, as the novel is for the one who stays up all night reading, than the bewildering view of morning. We return to Merleau-Ponty, who states:

> The temporal perspective with its confusion of what is far removed in time, and that sort of 'shrinkage' of the past with oblivion as its

[23] Proust, *Sodom and Gomorrah*, p. 499; F, III: 499.
[24] Proust, *ibid*, p. 508; F, III, 508.

> ultimate limit, are not accidents of memory, and do not express the debasement into empirical existence of a consciousness of time theoretically all-embracing, but its initial ambiguity: to retain is to hold, but at a distance.[25]

The subject *is* time, and brings time into being as he or she lives life. The intentional arc of Marcel's enactment of desire is that which effects the transition of the future into the past and the revitalization of that past in present acts of memory. It doesn't take a shocking instance of involuntary memory, like the boot button event, or the resurrection of a detail long repressed, like a sexual scene once glimpsed, to confirm that we are each a crossroads of time. This ambiguous position points to our freedom: insofar as we act according to the future coming towards us, we are free. Even our re-experiencing of the past stands as a new event in our efforts to make sense of it in this context, "unfolding it anew according to its own *tempo*."[26] At the end of *Sodom and Gomorrah*, Marcel refuses this possibility, and holds rigidly to a past narrative that requires self-punishment and denial of his homosexual desires. His grief is thorough: he alternates between repetition (mother as grandmother) to stop time and anxious awareness of the way the world changes around him. The end of volume four of *À la recherche du temps perdu* is bleak. Marcel achieves his self-punishing wish, a dark look from his mother (the look she wore on the night she last slept in his room, the look "which at this moment bore an extraordinary resemblance to that of my grandmother,") as he tells her, "I absolutely must marry Albertine."[27]

---

[25] Merleau-Ponty, *op cit,* p. 423; F, 483-84.
[26] *ibid.*
[27] Proust, *Sodom and Gomorrah*, p. 514; F, III: 515.

Lyrical Bodies: Music and the Extension of the Soul

Sander van Maas

Throughout its history, music has resisted subsumption under the prevalent categories of time and space. This resistance evidences, for example, in the development of musical notation. From the earliest beginnings musical notation has included concepts which, although referring to space and time as we know them, cannot be wholly understood within their confines. Examples of these are the gesturality of Gregorian chant notations, the notions of "rising" and "falling" common to our understanding of melody, and the "stacking" of polyphonic melodies in musical space. In contrast to these experiential metaphors of early practice, modern musical notation is supported by the notion that music unfolds against the background of a purely homogenerous temporality and spatiality. Music here appears as an object that calls for description in architectural and mereological terms, evoking as it does a combination of Aristotelean and Euclidean forms of space-time. The starting point of the exploration will be the notion of the sound body (*corps sonore*), which has been analyzed by Jean-Luc Nancy in terms of the soul. As van Maas argues in the second part of the chapter, the analysis of the soul is the key to understanding the new "liquid" or "lyrical" negotiations of space and time that currently take place in society (Zygmunt Bauman, Brian Massumi). This point will be illustrated with examples ranging from musical modernism (Olivier Messiaen, Harrison Birtwistle) to popular culture (Frank Sinatra, Miles Davis, Michael Jackson).

Listening: From Identity to Place

Where are we, when we listen to music? This question, which Peter Sloterdijk formulated in an essay from 1993, captures one of the key issues in contemporary music philosophy.[1] At first sight, it seems an odd question to ask, for the obvious answer is: within earshot of the music. In a concert hall, we are spatially close to the musical performers, and when we listen to recordings we are in the vicinity of loudspeakers. Even when we hear a song in our mind, we imagine ourselves to be close to that virtual music. So why, in the first place, is it important to ask the question of "where"?

As Sloterdijk writes in the philosophy of the 1990's, there has been a considerable increase in attention for the ear. This new interest, which has gone through another surge about a decade later, is according to him, an effect in the field of mediality and the senses, and of the discourse on the "end of metaphysics."[2] The decline of this key discipline, Sloterdijk argues, challenges the privilege it had accorded to visual phenomena and experiences.

---

[1] Sloterdijk, Peter, "*Wo sind wir, wenn wir Musik hören?*" in Peter Weibel (ed.), *Der Aesthetische Imperativ: Schriften zur Kunst* (Hamburg: Philo & Philo Fine Arts, 2007), p. 50-82. The question is a variation on Hannah Arendt's "where are we when we think?"

[2] The second surge of philosophies of the ear is recorded by such volumes as Szendy, Peter, (ed.), *Ecoute* (Paris: Ircam / Paris-Montréal: L'Harmattan, 2000), and Dell'Antonio, Andrew, (ed.), *Beyond Structural Listening? Postmodern Modes of Hearing* (Berkeley: University of California Press, 2004).

Now that philosophy is moving away from such classical metaphysical themes as light, contemplation, inspection, and form (*eidos*), it needs new concepts, models and figures to articulate its thinking. One source for this renewal is the domain of the aural.

Sloterdijk claims that the aural implies an epistemology which differs from the visual in that it involves a different spatial distribution of the knower and the object(s) of knowledge. According to him,

> The seeing subject stands on "the edge" of the world like a world- and bodiless eye in front of a panorama - Olympic contemplation and optical theology are just two sides of the same coin. By contrast, thinkers who understand being (*Dasein*) on the basis of the facts of hearing, would never have considered the removal of the spectator-subject toward the imaginary outer limits of the world because it belongs to the nature of hearing to arise no different than in the mode of being-in-sound. No hearing person can believe to be standing at the edge of hearable things. The ear knows no opposite position, it does not develop a frontal "view" of removed objects, for it has "world" or "objects" only insofar as it is among the acoustic events - one could also say: insofar as it hovers or dips in auditory space.[3]

The philosophical turn toward the ear, then, implies a spatial redistribution and redefinition that ultimately affects the whole of our understanding of the world, the subjects that live in it, and the way world and subjects interrelate. This interrelation Sloterdijk articulates by means of two diametrically opposed subject positions. On one hand, he envisages a "deep acoustic" according to which the subject may choose to radically withdraw from its dialectical dealing with the world by completely shutting itself off. This acoustic "sunkenness" (*Versunkenheit*), which he also refers to as *enstasis* (as opposed to ecstasy), is in itself a vital concept for our understanding of both inspiration (*Beseelung*) and "being-in-itself" (*in-sich-Sein*), Sloterdijk writes.

On the other hand, the acoustic realm offers a way to understand subjectivity according to a profound trembling, that is, according to a "deep psychology" that takes an extension of the soul as its point of departure.[4] Locating the origin of this second approach in the early nineteenth century, Sloterdijk contends that "Hegel for the first time articulates the idea that a still completely empty [unborn child's] soul that is devoid of experiences and passions and that therefore is undetermined, is penetrated in a decisive and

[3] Sloterdijk, "Wo sind wir," *op cit*, p. 52.

[4] Hegel, G.W.F., *Philosophy of Mind: Part Three of the Encyclopedia of the Philosophical Sciences*, trans. by W. Wallace and A.V. Miller (Oxford: Clarendon Press, 1971), p. 94.

shaping manner by the vibrations of the maternal medium."[5] Although Hegel's theses do not materially address the domain of music, Sloterdijk argues that they do in fact inaugurate a departure from the view of the soul as an impenetrable entity, and they introduce new ways to conceive of the interrelations between subjects (the unborn child and the mother, in this case) and the world.

The question: "Where are we, when we listen to music?" thus leads toward reconceptualizations of the boundaries of hearing (and, for that matter, listening) subjectivity. If indeed the subject is stretched between, on one hand the "in-itself" of a radical withdrawal, that is, a constitutive aural self-touching and; on the other hand, an equally radical *Durchzitterung* or trembling by forces, and according to an extension, that precede and penetrate it, the critical issue then becomes how these extreme insides and outsides are mediated. Where are we when, listening to music, we are neither *a)* listening across a (metaphysical-visual) divide between subjects and objects, nor *b)* wrapped in the self-touching aural sunkenness associated with the experience of the soul, nor *c)* reduced to the restless trembling and unlimited expansion of an aural vibration? Finally, on a more empirical level, this also leads to the question: "Where are we when we, listening to music, nevertheless continue to have the definite impression that we are touched inwardly by that marvelous music 'out there'?"

The aim of this essay is to map the space that unfolds in musical experience. The basic intuition is that, in course of the twentieth century, the question *where* I am has gradually replaced the question of identity, of *who* I am, which has dominated much of nineteenth century musical thinking. In the latter period the question of identity was often answered in terms of time. Hegel, for instance, explains the effects of music on the soul by drawing a close analogy between the temporal constitution of subjective inwardness and the movement of melodies and harmonies.[6] As I will attempt to show, the question of "the who?"; that is, of the soul of the listener, is only intensified by the music's folding, as a spatial and exterior phenomenon, into the constitution of this inward self. As will become increasingly clear, being touched by music is an example of what Jacques Derrida, referring to the thinking of Jean-Luc Nancy, has called the experience of "self-touching the other."[7] The space of listening will be approached as it unfolds in musicalized, lyrical bodies, which seem to touch one another without apparent contact. The lyrical bodies, then, obey the musical haptics that Plato described when he said that musical harmonies can "take strongest hold upon the soul" and

[5] Sloterdijk, "Wo sind wir," *op cit,* p. 72-73.

[6] Hegel, G.W.F., *Aesthetics: Lectures on Fine Art*, trans. T.M. Knox (Oxford: Oxford University Press, 1988), vol. II, pp. 906-908.

[7] Derrida, Jacques, *On Touching - Jean-Luc Nancy* (Stanford: Stanford University Press, 2005), pp. 265 *ff.*

that they indeed can "touch it strongly."[8] As I will argue, this soul, which in Plato remains purely ideal, will need to be re-thought according to the coordinates of a vibrational and extended *corps sonore* or, as Sloterdijk has it, a *medium percussum*.[9]

Extensity: "I am those tunes"

In spite of its ring of obsolescence, the concept of soul remains important for understanding musical experience, if only because it is still frequently used as a term. In realm of music the notion of soul is called upon, as a matter of course, by those dealing with it as listeners, players, or writers. Good music, we learn from our fellow listeners, "touches the soul." We frequently hear musicians voice their intention to play "from their soul." Music reviewers often appreciate music for having "soul." Indeed, there is a whole family of musical genres referred to as "soul" music. No word appears more apt for the expression of the value and importance of music - but no word from our musical vocabulary is less scrutinized either.

The persistence of the notion of soul suggests that the romantic aesthetic tradition that contemplated the soul, the heart and the inner sense (*Gemüt*) is still lingering in our everyday, "trivial romantic" idiom for describing musical experiences.[10] The meaning of the term often evokes the work of such poets as Charles Baudelaire, who used it to describe his intense listening experiences with Wagner's *Tannhäuser*. Interpreting Baudelaire's musings on music and poetry, Philippe Lacoue-Labarthe in his book on Wagner, highlights the poetics of lyricism to which both artists, albeit in different ways, subscribe. The soul, it appears, is found whenever experience becomes lyrical.

According to Lacoue-Labarthe, the lyrical experience is "the very essence of *transcendence*, in the active sense of the term, or, if one prefers, the *metaphysical* experience par excellence."[11] A typical example of the lyrical experience can be found in Baudelaire's description of it in an essay on the poetry of his contemporary Théodore de Banville, whose frequent use of the word "lyre" inspires Baudelaire the following comment.

> There is a lyrical manner of sensing. The men most disgraced by nature, those to whom fortune gives the least leisure, have

---

[8] Plato, *Republic*, trans. Paul Shorey, in *Plato: The Collected Dialogues*, edited by E. Hamilton and H. Cairns (Princeton: Princeton University Press, 1996), (401d), p. 646.

[9] Sloterdijk, "Wo sind wir," *op cit*, p. 78.

[10] See Bonds, Mark Evan, "Idealism and the Aesthetics of Instrumental Music at the Turn of the Nineteenth Century," *Journal of the American Musicological Society*, Vol. 50, Nr. 2/3 (Summer - Autumn, 1997), pp. 387-420. Carl Dahlhaus, *Klassische und Romantische Musikästhetik* (Laaber: Laaber Verlag, 1988), pp. 186-197.

[11] Lacoue-Labarthe, Philippe, *Musica Ficta (Figures of Wagner)* (Stanford: Stanford University Press, 1994), p. 25.

> sometimes known these sorts of impressions, so rich that the soul is as if illuminated by them, so lively that it is as if lifted up by them. The whole inner being, in these marvelous instants, launches itself into the air with too much lightness and dilatation, as if to attain a higher region. Thus there also necessarily exists a lyrical manner of speaking, and a lyrical world, a lyrical atmosphere, landscapes, men, women, animals, which all participate in the personality affected by the Lyre.[12]

Many listeners today, too, whose lives are often more leisurely than the men "disgraced by nature" referred to by Baudelaire, will have known this kind of experience, too. It even appears that in the globalized music business that has developed since Baudelaire's days, lyrical experiences aroused by music have become the most common thing on offer, constituting as they do the ultimate selling point for musical recordings and events. With the globalization of music, the claim to universality that had already been inherent in nineteenth-century descriptions of lyrical experience, has become stronger and more concrete. Through the globalized market listeners around the world are introduced to concepts of inwardness and musical ecstasy that may not have been part of their local interpretations of how human individuality is structured. Not even in our own cultural zone, may listeners be familiar with the precise implications of these historical usages of the word, "soul".

As a concept, then, the "soul" should be expected to betray its own singularity, both historically and geographically. As Jean-Luc Nancy would say, there is no "the" soul. On one hand, many cultures know words that are suggestive of musical experiences of (the) soul - a fact that arguably facilitates the process of its universalization. On the other hand, however, it has been stressed by various authors that the unequivocal translation of soul-words into foreign languages is difficult if not impossible, and therefore should be avoided. The fact that the neglect of trans-cultural differences can easily obscure the conversation is evidenced for example by Jacques Derrida's stage performance with the band of Ornette Coleman. In 1997, in the French capital, Derrida presented on stage fragments of a text that he wrote at the invitation of the saxophone player and band leader.

In his text, Derrida develops a line quoted from a conversation he had with Coleman in preparation for his performance. In this line, Coleman's mother is mentioned saying to her son, who at that point in his life was only starting his career as a musician and decided to play, in dubious circumstances, for white audiences: "Do you want to be paid for your soul?

---

[12] Baudelaire, Charles, *Oeuvres complètes* (Paris: Gallimard, 1976), vol. 2, p. 164; quoted in Lacoue-Labarthe, *Musica Ficta*, pp. 24-25.

That's what you want?"[13] In the ensuing elaboration of this phrase Derrida translates the English word soul with the French *âme* and gives it a conceptual structure typical of his own thinking. The soul referred to by mother Coleman, he suggest, is structured like an event.

> This is what his mother has said to him: one is not paid, one is not to get paid for one's soul, and the soul, is this, it is what happens, it is the music one is supposed to listen to and make - and which one should write: that is why one should not be paid because it is incalculable and priceless. The event.[14]

As Derrida was later to comment, the incalculability of the event could not have been better illustrated by the course of the Coleman concert itself. Soon after the beginning of his vocal performance, Derrida was booed by the hostile audience and he was soon forced to leave the stage. Although only few people will have actually listened to Derrida's words, his understanding of the notion of African-American soul carried the seeds of a cross-cultural confrontation. As David Wills comments,

> What is perhaps lost in the French translation of "soul", or at least remains unmarked in all the published texts that refer to it, is the connotation we get from the word thanks to African American vernacular. It is perhaps because of that that Derrida's text was adestined in its central modulation, as it seems to pass directly from "soul" to *âme* without acknowledging the ethnomusicological significance of the English word. That does not necessarily mean either that Coleman's mother was using the word in that sense back in 1947 (if I understand the chronology correctly) - although "soul" already suggested "pure human spirit" and "true American music" as early as Du Bois [*The Souls of Black Folk*, 16] - or that Derrida is unaware at least of the use of the word in the expression "soul music". for French uses English to refer to that musical style. Rather, as pure blackness within spiritual essence, or pure expression (somewhere on the fringes of jazz and gospel) within the repertoire of musical inspiration, "soul" functions as a type of untapped and abyssal reserve, a form of priceless musical event

---

[13] Derrida, Jacques, "*joue - le prénom*," in *Les Inrockuptibles*, No. 115 (September 1997), p. 42.

[14] "*Voilà ce que lui a dit sa mère: on n'est pas payé, on n'a pas à être payé pour son âme, et l'âme, c'est ça, c'est ce qui arrive, c'est ça la musique qu'il faut entendre et qu'il faut faire - et qu'il faut écrire: ce pourquoi on n'a pas à être payé parce que c'est incalculable et sans prix. L'événement.*" Derrida, *ibid*, p. 42.

> that inhabits and haunts the word beyond the way it is set in play or improvised upon in the texts in question.[15]

The soul, then, becomes a term that, rather than uniting the globe in a promise of collective and cross-cultural elation, marks out the unfathomable rifts that may exist between various ethnic cultures and, for that matter, between regions and nation states.[16] A case in point of the topographical singularity of (the) soul is the Turkish notion of *hüzün*, which expresses a particular, melancholic mood that, according to novelist Orhan Pamuk, typically and exclusively expresses the soul of Istanbul. The feeling of *hüzün*, as symbolic marker of the city, can be evoked outside of Istanbul, too, in particular by means of music and poetry. The key difference between *hüzün* - "a cultural concept conveying worldly failure, listlessness and spiritual suffering" - and the common melancholy described by authors from the West, such as Balzac, is according to Pamuk the collective nature of the former.[17] This collectiveness includes the stones in the pavement, the presence of the once glorious past, and elements from everyday scenes of life in Istanbul.

These scenes do not just remain a background for the inhabitants but determine their very destiny by breaking their will. "The *hüzün* of Istanbul suggests nothing of an individual standing against society; on the contrary, it suggests an erosion of the will to stand against the mores and values of the community, encourages us to be content with little, honoring the virtues of harmony, uniformity, humility."[18] *Hüzün,* then, is a notion that refers to a sense of identity that may be felt by an individual, but only in relation to a collective feeling (if such a thing exists) that occasionally acquires an almost objective quality. "On cold winter mornings, when the sun suddenly falls on the Bosphorus and that faint vapour begins to rise from the surface, the *hüzün* is so dense that you can almost touch it, almost see it spread like a film over its people and its landscapes."[19]

The aspects of the soul discussed so far - i.e., the body parts, the lyrical inwardness, the ethnic singularity and the topography of the collective soul - all converge in the typical experience that shall be the starting point for the following philosophical discussion of the term. This typical experience is exemplarily described by a user of the video website **You Tube**. In response to images from George Clinton's band Parliament Funkadelic, this user recalls what this music has meant to him.

---

[15] Wills, David, "Notes Toward a Requiem, Or, The Music of Memory," in *Mosaic*, special issue *After Derrida*, Volume 39, Number 3, September 2006.

[16] Famous expressions of national soul include the *deutsche Seele* (German soul) and the *velikaya russkaya dusha* (the great Russian soul).

[17] Pamuk, Orhan, *Istanbul: Memories and the City* (London: Faber and Faber, 2005), p. 82.

[18] *ibid*, p. 94. It should be noted that the *hüzün* of Istanbul is similar to the melancholy of *saudade* that is associated with Portugese *fado* and some Brazilian music.

[19] *ibid*, p. 89.

> This is music for and from my soul! I AM those tunes! They may have been fresher, better, more original in those days - no doubt about that! But when I've experienced this family two years ago, they still blew my mind, literally! I've had unique musical orgasms for four or five times during the six (!) hour concert, once lasting for about 15 minutes, having me dancing closed eyed in pure ecstasy!! Best concert, ever!!! Pure inspiration!!![20]

The concerts of Parliament Funkadelic are (still) known for their extreme duration and for the neo-tribal qualities associated with both the performers and the fan community. The ecstasy here described displays the key features of the individual lyrical-metaphysical experience, namely elevation, syncope ("blew my mind"), and *extensity*. The latter neologism denotes a fuzziness of the subject's boundaries regarding the piece of music that, while remaining an exterior phenomenon, only appears to increase the intensity of the experience. The ecstasy further references the neo-mythological African-American identity typically associated with this group, a sense of overarching community ("experiencing a family"), and the creation of a singular sense of place. The 1970's Parliament Funkadelic concerts usually involved stage props, such as a spaceship and other means to situate the tribe in an interstellar and eschatologically rhythmed time-space.[21]

In summary, the current usage of the notion of "soul" in relation to music should neither be confused with, nor reduced to, its meaning in the metaphysical and theological tradition. Claiming that one *has* a soul that is touched in particular musical circumstances is arguably different from the meta-physico-theological belief that one essentially *is* a soul. After the "end of metaphysics", the re-use of notions from this tradition will often need to be reinterpreted according to changes in the economical, social, cultural and philosophical contexts.[22] The question then remains what the notion of soul means beyond its colloquial usage and, by implication, why this notion has not yet been replaced by more recently developed (*e.g.*, scientific) vocabularies. In order to answer these questions, it will first of all be necessary to reinterpret the structure of the experience of (the) soul.

---

[20] Comment posted in September 2007 by YouTube user Tressco on a Parliament Funkadelic live concert registration video posted by user DreamersBlues, identified by other users as being of a 1976 concert at The Summit in Houston. At the internet, www.youtube.com (page accessed September 22, 2008).

[21] The relevant track is called "Mothership Connection." For Parliament's music and mythology see Danielsen, Anne, *Pleasure and Presence: The Funk Grooves of James Brown and Parliament* (Middletown, Connecticut: Wesleyan University Press, 2006).

[22] See de Vries, Hent and Weber, Samuel (eds.), *Religion and Media* (Stanford: Stanford University Press, 2001); de Vries, Hent (ed.), *Religion: Beyond a Concept* (New York: Fordham University Press, 2007).

The Extension of the Soul

What is a soul? According to the logic inherited from the Western metaphysical and theological tradition, the soul is something that one has. It is not something one *is*. More precisely, the soul is something one seems *to be* because one *has* it. In this respect, the soul resembles the body. A body is what I seem to be, because I have one. How do I know that I have a body? I know because I see it and can touch it. The body shows itself to me, or, at least, most of it does. And, most of it I can touch. When nothing is in the way it touches me, too, and it seems to touch itself continuously, in itself - and away from me.

Regarding the soul things seem no different. According to Hegel, the soul reveals itself in the mirror of the eyes.[23] The degrees of reflection I can follow with precision in the brilliance and dullness of their surface. Also the touching, now intense then imperceptible, of the soul moves me inwardly. The soul appears to be a place where I am closest to myself, in myself - but also away from myself.

As Jean-Luc Nancy has proposed, it will be possible to understand the distance between the body and the soul - which is often said to have become unbridgeable since Descartes - in terms of the distance that allows the body to be the body, and the soul the soul. This distance would at once keep them apart and function as their *trait d'union*, that is, articulating their non-identical identity.

A body, Nancy argues, should first of all be distinguished from a mass. A mass, which may be either spiritual or material, is a substance that is only penetrated by itself and in itself. "We often tend to think that the body is a substance, that the body is substance.... Substance has no extension whatsoever. The true idea of substance is not even a stone, but the *point*, which has no dimension."[24] A mass, Nancy points out, is "the self that is unto itself, without extension, which also means that it is without *exposition*."[25] It follows, then, that in the metaphysical tradition, despite the conventional representation of the soul as a thing, for instance as a little figure or angel exiting from the mouth at the instant of death, the soul was primarily thought of as a mass, a pointed essence.

The body, however, is extended. Regarding this proposition, Nancy and the tradition appear to be completely in agreement, but this is only seemingly the case. "If the body isn't mass, if it isn't closed in on itself and penetrated by itself, it's outside itself. It is being outside itself [*hors de soi*]."[26] He then goes on to re-think this exteriority in terms of the soul. "And

---

[23] Hegel, *Aesthetics*, vol. I, *op cit*, p. 153.

[24] Nancy, Jean-Luc, "On the Soul," in *Corpus* (New York: Fordham University Press, 2008), p. 123. Translation amended.

[25] *ibid*, p. 124.

[26] *ibid*, p. 126.

this is what is at stake in the word soul. It in no way involves an understanding of the ineffable interiority behind this word, a sublime or vaporous identity escaping from the prison of the body."[27] The soul is something the body *has*, and which at the same time it seems to *be*: a being outside of itself. It is the distance between the body and itself; that is, that which distinguishes it from a mass, or, for that matter, from a cadaver. Once the body loses its distance from itself, once it is concentrated, it becomes a corpse.

The answer to the question of the soul, then, leads Nancy from forms of concentration (mass, ego, corpse) toward notions of distance, extension, and ex-position. In a similar idiom one could say that the soul is the exile of the body: my body appears to me, and it allows itself to be seen and to be touched. More precisely: the soul is the outside *itself* of the extension or the ex-position. "The soul is the fact that a body exists, in other words, that there is extension and ex-position. It is therefore offered, presented open to the outside. A body touches on the outside, but at the same time…it touches itself as outside. A body accedes to itself as outside."[28] *That there is* outside: that is the soul.

In conclusion, for Nancy, the soul refers to the penetrability, the openness of the body. It refers to something "extra" a body has, or is, relative to the corpse. It is "extra" in the sense of exterior, and extraordinary, without having recourse to the sovereignty of the soul known from the metaphysical and theological tradition.[29]

Diapason: The Self as a Tone

Is the soul silent? The ways in which the soul becomes manifest are often described in terms of inwardness. Inside of me, I experience my soul as a vital force, as a sensibility, as a power to relate and to love. According to Roland Barthes, writing about the music of that particular age, "feeling" and "heart" were romantic names for the body.[30] Just like "soul" they shed light in its inner darkness but they were also indicative of its silence. The soul, Nancy writes, is the silence of organs. "Only in silence is there true intimacy of the body."[31] Hence, if lyricists are right in claiming, as they often do, that the soul can sing, this must be a silent singing. That is, a singing that takes place in the rest - or fermata - of the body.

---

[27] *ibid.*

[28] *ibid*, p. 128.

[29] Insisting as he does on the extension of the soul, Nancy introduces a *glissement* between two apparently incommensurable forms of space or "two spacings." See Derrida's comments in *On Touching*, p. 24.

[30] Roland Barthes, "Rasch," in *The Responsibility of Forms: Critical Essays on Music, Art, and Representation* (Berkeley: University of California Press, 1991), p. 308.

[31] *ibid*, p. 118.

What is the rest of the body? Nancy introduces "a rest" between the body and itself. This rest he also refers to as an interval, which is the appropriate term in the present context. Obviously, the body is not silent. Sound artist Henry Chopin called the body a sound factory and exploited its continuous production of sounds in his sound poems. In *Chant du corps*, for instance, he recreated the body electronically on the basis of its most intimate sonorizations.[32] The Dutch poet Martin Reints remembers the first nights after he moved from the city to the countryside of Friesland. "I almost forgot, but during the first nights I was amazed by the sounds produced by my own body. It bubbles and gurgles. You hear your ears ring. In [the urban area of] the Randstad these sounds were completely drowned out."[33]

The interval Nancy refers to is not some melodic or harmonic proportion within the song of the body. Is it not a strange harmony arising from the bubbling and the gurgling. Rather, this interval refers to the possibility *as such* of this song to resound. Indeed, the song of the body that takes place in its "rest", in the soul. The soul, being as it were, the distance between the body and itself, *has* music. And perhaps it *is* music - or rather it *appears* to be music because it *has* a certain musicality. Of course this is a strange way of speaking. To "have music," or even to "be music": things like these one cannot say. Arguably, this is no coincidence. Music is the resistance, as such, against "having" and "being," that is, against the principal verbs of classical metaphysical thought.[34]

Perhaps this occurrence explains why it is so natural to use the vocabulary of the soul, and, in spite of historical changes, to keep using to it. No more than I have a body, or than I am a body, do I have a soul, or am I a soul. Indeed, the reverse would be easier to accept. Music inserts itself in the interspace - the interval - between my body and myself. As mentioned before, this interval is not an interval between two pre-constituted entities. The interval of the body, that is, the interval which is the soul, is more originary than all other entities.

The logic here is strictly musical, for in music, the interval is that which precedes the individual tone. Without the distance between a tone and itself, without the auto-resonance that multiplies a given vibration to produce a series of overtones, there would be no timbre. Without timbre, tones cannot individualize by acquiring a proper character and will be destined to remain musical raw material. In recent music history, the insight into the necessary reversal of the hierarchy between tones and timbres has given birth to so-

---

[32] Chopin, Henri, *Chant du corps* (1966), i.e., Part III of *Le corps*, at http://www.ubu.com/sound/chopin.html.

[33] "Martin Reints: De ballade van de winstwaarschuwing," in *De Groene Amsterdammer*, Magazine special edition for De VSB Poëzieprijs 2006.

[34] In metaphysics, "being" pertains to substance and having qualities that can be predicated of a substance.

called "musical spectralism."[35] In the 1970',s the composers Gérard Grisey and Tristan Murail, among others, have drawn creative consequences from their understanding that the interval is the distance, the suspension, and also the extension that is presupposed by every sound. Without the interval, no harmonic series would be possible, let alone tones and intervals between tones.

The Soul and Music

Between the soul and music, there is neither *having* nor *being*. There is only an in-between, an interval, where the concepts of soul and music resonate. Nancy thinks "after metaphysics" and "after religion." His work is an attempt to give up thinking of the soul as a metaphysical or for that matter theological substance, which serves as the pointed essence, or as the focal point of pure being. For this notion of the soul, Nancy substitutes a interpretation of the subject that appears to be based on the logic of musical resonance. Nancy calls it the "diapason-subject." This turn of thought, it should be added in comment, is risky. *Diapason* is an old word referring to the interval of the octave, and in French is still used to refer to the tuning fork and to the tone played by the oboe when an orchestra needs to tune. By its name, then, the diapason-subject refers to two of the more powerful means that have helped to turn music into a system or a regime.

First, the octave is the preferred object of music theoretical ideologies that insists upon the naturalness of the tonal system that is often associated with the dominance of the tonic. Second, the tuning fork has been the preferred means to introduce in music absolute pitch, namely by declaring the pitch "a" at 440 Hertz the point of reference for all tonal systems. For the last time in history this absolute pitch was established in 1939 at an international conference in London at the expense of other tunings in use until then. [36] The diapason-subject, then, appears to invite the logic of universality and identity that had been foundational for the metaphysics of substance that Nancy intends to leave behind.

This fact, however, is not Nancy's own interpretation. The diapason-subject captures the structure of a particular mode of being, a type of presence that, as Peter Sloterdijk writes in the passage quoted in the introduction above, differs from specular forms of presence. When I am listening (*être à l'écoute*), I am "there" but in an unusual manner: it is as though presence itself is doubled against itself.

[35] Jean-Luc Nancy refers to this compositional current that sprang from students of Olivier Messiaen in *Listening*, pp. 39-43. On "spectralism", see the French volume *Barrière*, Jean-Baptiste (ed.), *Le Timbre: Métaphore pour la composition* (Paris: Ircam/Bourgois, 1991).

[36] Historical performance practices have counteracted this norm by reintroducing ancient diapasons and tuning systems. These notwithstanding, the dominance of the a = 440 Hz remains firmly in place today.

> [Sonorous presence is] always within return and encounter. It *returns* (refers) to *itself*, it *encounters* itself or, better, occurs against itself, both in opposition to and next to itself. It is co-presence or, again, "presence in presence," if one can say that. But insofar as it does not consist in a being-present-there, in a stable, fixed being, yet is not elsewhere or absent, it is rather in the rebound of "there" or in its setting in motion, which makes it, the sonours place ("sonorized," one is tempted to say, plugged into sound), a place-of-its-own-self, a place *as* relation to self, as the taking-place of a self, a vibrant place as the diapason of a subject, or better, as a diapason-subject. (The subject, a diapason? Each subject, a differently tuned diapason? Tuned to self - but without known frequency?).[37]

Without a known frequency: as a subject I am a proportion with myself, tuned (in) on myself, and after the model of myself. However, since this proportion is an interval, I am never simply given to myself. I remain a vibration, a trembling, or shivering, which goes deeper than whichever grounding tone.[38] For the outplace of my soul condemns me to the life of minstrel, making me search forever for my *Eigenfrequenz* - for the optimal situation in which I look into the acoustic mirror that gives me to myself, and hence to the world.

The diapason-subject is, in a sense, a musical instrument and as such it evokes the dialogues between Socrates and the Pythagoreans. In these, Socrates aims to show that the soul is immortal and will survive death of the body. Countering this claim, Simmias, the Pythagorean, defends his view that the soul is finite by drawing on an analogy between the soul and a lyre. He argues that, much in the way the parts of a lyre are harmoniously attuned by the tuning of the instrument, the soul is a harmony of the various parts of the human body. When these parts perish in the advent of bodily death, Simmias counters his interlocutor's claim, their harmonious tuning will also perish. End of argumentation. The metaphor of tuning the lyre used in this argument suggests that the question of the finitude of the soul ultimately becomes a question of proportion. For the *Eigenfrequenz* of a string, that is, the typical frequency that results from the string, is a function of its mass, its length (*mensura, diapason*) and its tension. The diapason-subject is determined by the proportion between these factors.

The extension of the body - in other words, the soul - is an interplay between mass, spatialization, and tension. "The body," Nancy writes plainly,

---

[37] Nancy, *Listening*, *op cit*, pp. 16-17. On the genesis of the place (*le là*) see also Nancy, "On the Soul," pp. 132-33.

[38] On the Hegelian notion of trembling, which resonates in Nancy's discussion of sonorous presence, see Sloterdijk, "Wo sind wir," pp. 72-76 and Nancy, Jean-Luc, *Hegel: The Restlessness of the Negative* (Minneapolis - London: University of Minnesota Press, 2002), p. 44.

"is a tone."[39] This tone, however, is not a phenomenal tone: I cannot hear it. Its unique frequency remains unknown. The body is not a *corps sonore*, it does not coincide with its bubbling and gurgling. The body *is* not this music and it does not *have* this music either. There remains an interval between, on one hand, the acoustic life of the body, and of the soul on the other. The interval is a resistance that distinguishes the body from a mere mass, and which makes acoustic life possible in the first place. This life exists on the basis of a deferral - an interval or, in the temporal domain, a *syncope*.

Goose Bumps

Whence comes the music of the soul? If this music is its own syncope, deferring as it does the interval by means of the interval, infinitely displacing itself toward a quasi-transcendental position, it will be difficult to localize it in the empirical domain. The music of the soul resides neither in a fixed place, nor (re-)sounds in its proper moment. Rather than expressing itself by means of a composition, it constantly defers itself in a kind of disposition. What is more, the instruments of its production are not real - such as this lyre here, or the voice of that body over there - but virtual. The music of the soul is rather produced, it may be said, in a contact between lyres and voices and bodies, that is: in the world.

The soul will often make me grab my heart but not because my heart misses a beat. On the contrary, the soul is the experience of opening oneself to the world, and as the world - it is an experience of bleeding.[40] Would it be possible to bleed, but only to a moderate degree? That is what many definitions of music seem to demand. Augustine, for instance, defines music as the *scientia bene modulandi*, the science of good measuring. A soul that bleeds, then, shall have to modulate in its opening to the world. This means that the flux of bleeding will have to be stemmed by a bar or measure. The measure is to be found in the *modulation* of my diapason by the music that I hear, and with which I (co-)constitute an interval.

Music modulates the diapason of the subject. I do not appear to myself in the immediacy of my self-experience. Rather, I appear to myself in the modulation of my diapason, and *as* the modulation of my diapason. That is, the music I hear does not leave my subjectivity untouched, it tunes and de-tunes the proportion I have to myself. This proportion is a relationship within the world, in the exteriority that is called my soul.

> Isn't the space of the listening body…just such a hollow column over which a skin is stretched, but also from which the opening of a mouth can resume and revive resonance? A blow from outside, clamor from within, this sonorous, sonorized body undertakes a

---

[39] Nancy, "On the Soul," *op cit*, p. 134. Translation amended.

[40] See the discussion of Pamuk's *hüzün* above.

> simultaneous listening to a "self" and to a "world" that are both in resonance.[41]

The soul is not the pointed essence that is to be thought as the innermost essence of the listener. It is the space of a resonance, a reverberation, an echo. The soul is extended, it touches right on (*à même*) itself and on the world. It is like a *timpanum* that animates either of these domains simultaneously: "self" and "world." It is the *touché* with the music "out there" through which the self - that is, the diapason of my subjectivity - is made to resonate. It is also the beat, or vocalization, that makes me find my world, in and through which I experience myself as soul in the music out there. The soul is the interval in-between these domains that resonates in either direction. It is neither my inner limit as a pointed essence nor the exterior border constituted by the musical "object" that I listen to. It is an intimate exteriority, or - which is the same according to the logic of resonance - and exterior intimacy. "I AM those tunes."[42]

Resonance is a local effect. The experience called soul is never homogenous or guaranteed. It is divided, disposed into zones, into regions that are more or less resonant. Not all music modulates my diapason. The number of the modulation remains a singular number, even a secret number: I can only find myself as soul by wandering around, like a minstrel, knocking on sound bodies, vocalizing in the spaces that I enter, or tuning into my metronome.

Film maker Oskar Fischinger once said to John Cage: "Everything in the world has its own spirit, and this spirit becomes audible by setting it into vibration."[43] This inspired Cage to develop percussion music using world around him. "[Fischinger] started me on a path of exploration of the world around me which has never stopped - of hitting and stretching and scraping and rubbing everything."[44] This quest to listen in to the soul of things is echoed by Louise Bourgeois' use of a metronome to find her own rhythm. In a documentary she tells the interviewer, after painstakingly adjusting the tempo, "This is it. At this level we are in business. You can talk to me." [Asked about the use of the metronome, she explains], "It allows me some self-knowledge. I just put it on, and I find my rhythm."[45] This notion of finding oneself as a resonance with the world (or even in resonance as a world), is also found in the singing practice of Iégor Reznikoff. Singing his reconstructed Gregorian chants in such spaces as Paleolithic caves and Cistercian abbeys (Vézelay, Fontenay), he noticed how important it is to

---

[41] Nancy, *Listening, op cit,* pp. 42-43.

[42] See also van Maas, Sander, "Intimate Exteriorities: Inventing Religion Through Music," in de Vries, *Religion: Beyond a Concept*, *op cit,* pp. 750-771.

[43] Revill, David, *The Roaring Silence: John Cage - A Life* (New York: Arcade, 1992), p. 52.

[44] *ibid.*, p. 52.

[45] In *Louise Bourgeois*, director Nigel Finch (BBC, Broadcast, 1993).

stand exactly in the right place in order to experience a sonorous anamorphosis; that is, according to him, as much an anthropo-somatological as an acoustical experience.[46] I can only find myself as far as I regard myself as a diapason-subject, as a living tuning fork that must be held right in front of a loud speaker cone in order to detect a possible resonance. The world-as-soul opens itself according to place: a movement is needed to find the position that renews me. The soul and the world are always one, and always new. (If indeed they are there).

For often there are dead spots. Every sound body has them. They are, for instance, those places on the neck of a guitar that will not resonate to the vibrations of the strings - places where, locally, resonance will forever be extinguished. Similarly, there are those zones in my body that are irresponsive to the music out there - occasionally this concerns my whole body. Outside of resonance my body is a mere object, a mass, a cadaver. It is world-less, place-less, and merely an object in a homogenous space. A thing among things, it is then like a heart that has stopped bleeding, a resonance that has been extinguished, an interval that has imploded, or a syncope that has left its orbit. The end of resonance is the end of sense in the completion of sense: it turns the interval into a relation, such as, for instance, the distance that exists between music as an object and its listener as a subject. Here it would be necessary to distinguish, as does, for instance François Nicolas, between a resonant listening (*écoute*) and a objectifying hearing (*entendre*).[47]

Music appearing outside of its resonance is mono-medial, addressing the ear only, disengaging the act of listening from its modulating engagement with other senses. It is the music that I can always hear whenever I want to, and as often as I want to, as long as it is available. As Nicolas writes, his hearing resembles the inspection of a building: it surveys its parts and structures, it makes an inventory of the motives, registers the impression made by the sonorous totality. It does this by hypostatizing the interval, by distributing the roles and surfaces beforehand, and stabilizing their disposition. The listener here is a subject that perceives from a distance - the hypostatized interval - the music that offers itself objectively for inspection. On this side of this interval, there is a sympathetic inwardness (of the subject), on the other side (of the music), as well. This inwardness, however, will never be recognized as one's proper intimacy, as the sonorous surface called soul. Rather, they remain caught in a space-time that, as

---

[46] Reznikoff, Iégor, "On the Sound Dimension of Prehistoric Painted Caves and Rocks," in Tarasti, Eero (ed.), *Musical Signification: Essays in the Semiotic Theory and Analysis of Music* (Berlin - New York: Mouton de Gruyter, 1995), p. 544. On the concept of anamorphosis in Reznikoff's experiments, see Jean-Luc Marion in *Being Given: Toward A Phenomenology of Givenness* (Stanford: Stanford University Press, 2002), p. 123 and p. 350n6.

[47] Nicolas, François, "De l'aspect à l'inspect: La Forme musicale comme empreinte," on the internet, (page accessed September 24, 2008).
www.entretemps.asso.fr/Nicolas/TextesNic/Ulm.html

Heidegger would say, is set up and linearized according to a metaphysical reduction.[48] Is there a way to circumvent this unexpected reintroduction of a visual logic in contemporary listening practice? One final example suggests there indeed is.

If Heidegger's view of the reduction of space-time in metaphysics is read against the recent developments in music, one is lead to conclude that metaphysics, culminating as it does according to him, in techno-science, has turned the art of music against itself. The key example here would be the genre called *house music*, which since its breakthrough in the late 1980's, has spread rapidly to all areas of music creation.[49] The musical mechanicity celebrated in this music, and in many of its associated genres such as techno, hardcore, and *gabba*, only appears to confirm the idea that the intensification of technology in the West entails an aesthetics that encourages music to freeze into a metaphysical posture. However, the bodily experience of this type of (dance) music has shown the opposite to be the case. Commenting on the intertwinement of Promethean and Dionysian motives in techno, Michel Gaillot writes,

> It may...seem surprising, even contradictory, that the techno movement has been attracted to machines and technology, generally considered a power for separating and isolating individuals, a power for enslaving and slacking bodies. But it is precisely because of their potential to favor the free development of sociability and "world-ness," as well as of the body, that they have been claimed and promoted for techno, as if, far from separating singular existences, they might in fact contribute to commingling them more than political ideals could. It is not the least of the paradoxes surrounding techno that it is through the technological that the tendency to transgression - to phenomena of trance and ecstasy - survives and revives.[50]

Techno, then, unearths the relational powers of technics, creating a community that upholds the paradox of being at once infinite, in the sense that it re-creates the sacred, and ephemeral, inscribing itself as it does in the

---

[48] A parallel between music and architecture would be the difference described by Brian Massumi between the mereological view of buildings and their experience according to the topology of a biogram. See Massumi, Brian, "Strange Horizon: Buildings, Biograms, and the Body Topologic," in *Parables for the Virtual: Movement, Affect, Sensation* (North Carolina: Duke University Press, 2002), pp. 177-207.

[49] See Reynolds, Simon, *Generation Ecstasy: Into the World of Techno and Rave Culture* (Boston: Little, Brown, and Company, 1998).

[50] Gaillot, Michel, *Techno: An Artistic and Political Laboratory of the Present* (Paris: Dis Voir, 1998), p. 31.

festive suspension of time and space. In the rave, the diapason-subject is *durchzittert* by the bodily experience of bass sounds and, listening through the layers of endlessly modulating patterns, is given a sense of infinite aural and visual space.[51]

This more originary possibility - which is typical of, but by no means limited to techno - is the musical experience of the soul. It is the experience of an inter-medial type of listening, of a musical *touch*. The soul is a resonance along the surface of my body. It is the goose bumps that make me believe in a blissful (Platonic) penetration of my inmost, private inwardness by the subtleness of music. The suggestion of penetration, however, is only secondary with regard to the exteriority of the touching. The soul is not interior; it is an extreme exteriority. It is the ecstasy in which my body, my diapason, is modulated by music, and in which; conversely, music is modulated by my body.[52]

Wrapped in goose bumps I experience my body as music, and music as my soul. But even more than this intensity I experience the interval - large as the cosmos - which will forever keep me from coming home with myself, and as myself. The soul, the sound body, is resonance. It is listening oneself to death by an intensification of that which resists *having* and *being*: music, the interval, the syncope.

Epilogue: Music-touching the Self

Where are we, then, when we listen to music? Listening to music, as it has been suggested above, we are in a sonorous present. This "present" creates a place that remains wholly entangled with the music. Because the music - in such cases when it really becomes a *place* rather than a mere façade - is no objective circumstance; I cannot move around in it. The place created by music is *me* - that is, only "as long as the music lasts" (T.S. Eliot). This means that I can indeed walk around *with* it, as many people do when they listen through their mobile players.

The creation of this place is a singular event that cannot be foreseen. It happens according to the tuning process that I have attempted to describe, that is, by tuning in to the music available in our environment and by testing

---

[51] As Job de Wit describes in his book on the breakthrough of house music in the Netherlands, the audience were initially hostile toward this new type of music. Only after combining it with the drug Ecstasy did pioneering DJ's and clubbers understand the connection between the music and the type of experience. See de Wit, Job, *RoXY en de house revolutie* (Amsterdam: L.J. Veen, 2008), p. 64.

[52] The latter claim is supported by the fact that my readiness or refusal to accept the music, that is, my porosity, will affect the manner in which music diffuses in space. Nancy's comparison between a body and a bi-directionally working drum skin is even more accurate. It should be expected that my body broadcasts a singular "bodyprint" in manner similar to the ear's "otoacoustic emissions" described by medical scientists. See Wenxuan He, *et al.*, "Reverse Wave Propagation in the Cochlea," in *Proceedings of the National Academy of Sciences of the United States*, vol. 105, Nr. 7 (February 19, 2008), pp. 2729-2733.

its connection with "us" at that particular moment, in that particular spatial context. If there is connection, we may experience the advent of a new whole, a new sense of self, and consequently a new sense of the world. A new place is born and we tend to encircle it as though it were a sacred place (*temenos*).[53] This new place, which may indeed be nothing other than a *sense* of place (like the *hüzün* of Istanbul that can be experienced everywhere as long as the proper music plays), is best approached in topological terms. As Nancy shows in *Listening*, it typically deals with a folding of insides and outsides, a disposition of zones and intensities, and a play with our sense of contiguity.[54] This dynamic of connections and contacts (ensembles, *touchés*, beats, scratchings, rubbings) is what creates a continuity between the musician, her instruments, her hands, our bodies, my soul, and the musical sense that this all appears to have.

Yet, all through this series, there remains the logic of touch that not only operates a continuity but also presupposes an interruption. As Derrida is careful to point out in his sympathetic critique of Nancy, all touching implies a distance that cannot simply be overcome by adopting a theory of continuity.[55] The interruption between me and myself, between the listener that I believe I am - with all the history and hope that this notion implies - and the experience of being touched by music - this interruption arguably hides a desert that makes me despair over the impossibility and elusiveness of all that music promises.[56] The extensity and "extimacy" (Lacan) of the musically mediated experience of self strictly coincides with the logic of touch. The lyrical self is a detour, passing as it does by the music, and hence by all the other selves and listenings that this music carries and that is supported by, in an experience that may be called a "music-touching the self."

---

[53] See also van Maas, "Intimate Exteriorities," *op cit,* p. 768.

[54] Nancy, *Listening*, *op cit,* p. 38.

[55] Derrida, *On Touching*, *op cit,* pp. 124-25.

[56] See also van Maas, Sander, "Unfolding the Pocket, Neutralizing the Lyrical: Contrabass Guitarist Anthony Jackson and the Aesthetics of Modernism," in *Dutch Journal of Music Theory*, vol. 13, Nr. 3 (November, 2008).

Phased Space

Raviv Ganchrow

From advanced imaging devices to emerging spatial acoustic systems, developments in contemporary technology display a hybridization of sonic and optical attributes. Underpinning these various techniques are complex understandings of spatialized fluid dynamics. This paper traces implicit interdependencies emerging from within the confines of phased space, first as they appear in the research of the physicist / philosopher Ernst Mach, and later reflected in the contemporary application of wave field synthesis. Beginning with Mach's proto-spatial definition of "elements," a position is examined whereby both the "thing" and the discreet "ego" are viewed as alternating pattern-bundles composed of common blocks. The emphasized intertwining of observer and observed along with a broadened definition of 'sensation' is then shown to manifest in Mach's empirical research as a radical re-thinking of the epistemic purpose ascribed to the photographic plate. Numerous photographs, directly imaging "processes" (*e.g.* acoustic refraction, ballistic shock waves and turbulence), taken over the course of five years spanning 1888 – 1893 elucidate this position. In contrast to normative optic models, Mach arguably approaches the photographic process as an inanimate version of "sensation taking hold." Through this approach, the photosensitive emulsion is understood to be a subtle "memory substance" that engages and preserves unfolding "sensation" complexes. The theme of interdependent tactility and fluctuant spatiality is then applied to a reading of the wave field synthesis technique. This emerging cultural practice involving spatial sound rendering extends the "memory substance" modality into the audio spectrum by implementing a bounded acoustic expanse constructed as a vibrational tapestry, within which a spatialized act of listening occurs divorced from all source resonating "objects".

How can the pronounced temporality of sound be understood in terms of corresponding spatialities? For the situated listener, sounds seem to surround a point of observation, giving an impression of apprehension at a distance. Yet sound is also a physical phenomenon enveloping the listener. Sound communicates its contours by way of tactile intrusions upon the surface of the eardrum. In spatial terms, sound simultaneously occupies two locations for the listener – the one situated at a distance, the other positioned strikingly close at hand. Frequency oscillations are also distances traversed, to the extent that each cycle heard is also a spatial disposition with intervallic attributes that unfold in a space where tones can be seen to fit into a corresponding scale of physical sizes. The listener is at once immersed in such fields of fluctuating waves as well as located between discreetly oriented sounded events. An application of acoustics, considered in terms of vibrational "matter", can be discerned in auditory practices implementing phased array loudspeaker configurations. Wave Field Synthesis pronounces one such take on a materiality of sound where localized tactile interactions between "observer" and "observed" are simultaneously the posture from which immense soundscapes unfold.

This text aims to highlight certain aspects of sound by coupling considerations from wave field synthesis[1] with aspects of Ernst Mach's approach to sensations. The attempt is to set up a system of resonances that intentionally exceeds both contextual frameworks: on the one hand relating Mach's principles to practices that extend beyond his original investigations; and on the other hand, applying a theoretical framework to practical techniques of spatial sound production that do not outwardly announce such theoretical pretensions. This particular coupling is focused upon an account of what might be termed "Phased-space,"[2] a meta-perceptual domain of phenomena within which both Mach's experiments and wave field synthesis are seen to intervene. My intention is to outline the implications set forth by a materialization of "Phased-space", particularly with respect to the bearing it has upon the relations of sound-space-listener.

Ernst Mach recounts an autobiographical experience as the foundational moment for his intuitions of material existence:

> I have always felt it as a stroke of special good fortune, that early in life, at about the age of fifteen; I lighted, in the library of my father, on a copy of Kant's *Prolegomena to any Future Metaphysics.* The book made at the time a powerful and ineffaceable impression upon me, the like of which I never afterwards experienced in any of my philosophical reading. Some two or three years later the superfluity of the role played by "the thing in itself" abruptly dawned upon me. On a bright summer day in the open air, the world with my ego suddenly appeared to me as one coherent mass of sensations, only more strongly coherent in the ego. Although the actual working out of this thought did not

---

[1] My recent development of a wave filed synthesis system for electro-acoustic music in the Netherlands was designed to provide a 10 meter x 10 meter listening arena for acoustic rendering. The aesthetic-cultural implications of WFS as a platform for electro-acoustic music production were of primary interest in this project, particularly with regards to the social contexts and theoretical concerns of music and listening. The project, initiated by the Game of Life foundation, builds on experiments I conducted at the Institute of Sonology, Den Haag between 2002-2004. The initial algorithm (a geometric implementation of dynamic space-time mappings), has been extensively developed into a full GUI user interface by Wouter Snoei with the assistance of Jan Trützschler von Falkenstein. Generous technical support throughout the project was provided by Dr. Diemer de Vries and the Technische Natuurkunde department at TU Delft, the department in which the technique of wave field synthesis was originally pioneered.

[2] The use of the term "Phased-space" is to assign a nomenclature to the spatial sound field seen from the position of the wave interactions themselves. It is a category that encapsulates both space and sound in a single description. Phased-space should not be confused with the term "phase space" from mathematical analysis. It is my intention to emphasize the phenomenal underpinnings of phased-space.

> occur until a later period, yet this moment was decisive for my whole view.[3]

As a specific sensation, sound would become an important object of investigation in working out his original intuitions. Sound provides a pertinent subject matter in which to test correlations between the observer and the observed. Sound emanates from a realm of crisscrossing vibrations, though for the listener sound never appears in the shape of its vibrations. Every sound emerges in perception as a fluctuations (re)cognized. As a result, most profoundly, audible sounds are always chronometrically articulate as well as spatially present. Indeed, a sensation of a sound cannot be perceived that is not already emanating from a discernable source. Even with the case of deep bass tones, sound seems to emerge from "everywhere" and never from "nowhere". On the other hand, if we were able to peer into the expanse of vibrations, there would be little hope of identifying similarities between the spatio-temporal articulations familiar to our audible worlds when compared with the dense mass of enfolded sonic activity literally suspended in air.

The interplay between these two aspects of sound: the space of acoustic eventfulness, and the unfolding spaces of listening is the relational paradigm underlying the technique of wave field synthesis. Wave field synthesis (WFS) is a spatial audio rendering method that applies algorithmic calculations of acoustic transformations in the service of materializing unfolding fields of acoustic vibrations.[4] WFS utilizes individually amplified, sample synchronized, transducer arrays in order to dynamically shape acoustic vibrations into a choreography of spatial-sound relations. Within a four-sided array, arranged to provide a rectangular territory for listening, it becomes possible to materialize a confined field of sonic interactions cut out as it were from an enormous acoustic expanse.

Wave field rendering implements an understanding of acoustics that does *not* try to replicate hearing. It is a method that stands in sharp contrast to other electro-acoustic techniques that *simulate* the perception of localized sounds. Simulations will often employ physiological and psychological aspects of hearing (such as mapping the directional frequency characteristics of the pinna) to give an impression of sound localization in an imaginary

---

[3] Mach, Ernst, *The Analysis of Sensations,* (Chicago and London: The Open Court Publishing Company, 1914), p. 30.

[4] These combinations of calculations describe conditions where the "state" and "location" of energy is determinant within a given moment in time. What is essential for the spatial implementation of these calculations are "phase" relationships between adjacent sound rendering transducers. In acoustics, the term "phase" designates the particular state within the cyclic periodicity of compression and expansion. It is both the "state" and the "time" of an acoustic condition, and when extended to a line of transducers it is also the component responsible for the properties of curvature ascribed to synthesized acoustic waves.

space.[5] The result is a kind of duplication of aural functions back into sound thus imposing an "externalized ear" onto the matter of acoustic waves. In such audio setups, the sounds we perceive have already passed through a pair idealized ears.

WFS acts upon the air itself, in the domain of physical acoustics. Its operation relies on considering sound in terms of its material manifestation, and thinking of frequencies for their corresponding lengths and sizes. Instead of measuring sound from the gauge of an ear, WFS acquires its measure from the crest to crest spacing of successive traveling spherical waves. The technique maps vibration into spatial contours, shaping volumetric patterns of emission within a given confine of air. Phase-synchronized arrayed transducers imprint within the air multi-directional compression fronts, weaving in and out of one another to form a weightless tapestry of interpenetrating vibrations. The result is a time-space intervention into the pre-corporealized realm of pulsating matter. By coordinating the phase interactions on adjacently sounding loudspeaker membranes, one effectively multiplies the timbral characteristics of a sonic event with a spatial parameter thus enfolding both space-time and sonic-form into a single field of affect.

This technique impresses itself upon the Phased-space of fluid dynamics, in a mute territory of *pre-auditory* acoustic potentialities. The listener enters this remote terrain as a kind of experiential postscript. Immersed within vibrational interminglings, listening becomes an activating agent capable of unlocking the myriad of spaces enfolded within the continuum of vibrations. From the position of pulsating matter – there is no space to be discerned – the extended remains enfolded on a continuum. Any static location cast into that continuum acts as a ballast against the turbulence of acoustic flows, and every ear becomes a reference point for phases to brush up against and forms a locus through which durations take hold. It is only after consecutive phases have extended into durations that acoustics

[5] Particular techniques that implement HRTF (head-related transfer function) for binaural encoding embody the idea of a synthetic ear. Stereo, and arguably surround-sound systems, enforce another model of hearing namely a spatial paradigm fastened to an ocular analogy of space. By situating the listener in a precise relationship to two or more loudspeakers, one is able to give an impression of a soundscape that extends outwards from an apex at the ears and extends behind the loudspeakers. When properly situated, two loudspeakers construct a picture window for sounds, where distinct sound-objects are seen to interact in an imaginary ocular expanse behind the loudspeaker boxes. Even though hearing is a 360 degree affair in both the horizontal and the vertical axis, operating in all directions in every given instant, stereo decisively narrows that field in order to correspond to the frontally oriented apparatus of vision. Surround sound formats that have become commonplace in domestic settings, merely expand that frontality to include sounds-to-the-side or sounds-from-the-back. Peripheral sounds enforce the pronunciation of a distinctly axial understanding of the listening space, often culminating in a viewing screen or monitor marking the position of the "absolute front". The development of spatial sound technologies in concert with cinema standards (and their various domestic offshoots), underline the ocular basis of such sound models that aim to enhance or at least support the field of vision.

transform into the various dimensions of aural spatial sensations. In WFS sounding objects remain concealed entities until the participation of experience intervenes in the site of vibrations. As Mach writes:

> There is no rift between the psychical and the physical, no inside and outside, no "sensation" to which an external "thing", different from sensation, corresponds. There is but one kind of elements, out of which this supposed inside and outside are formed – elements that are themselves inside or outside, according to the aspect of in which, for the time being, they are viewed. The world of sense belongs both to the physical and the psychical domain alike.[6]

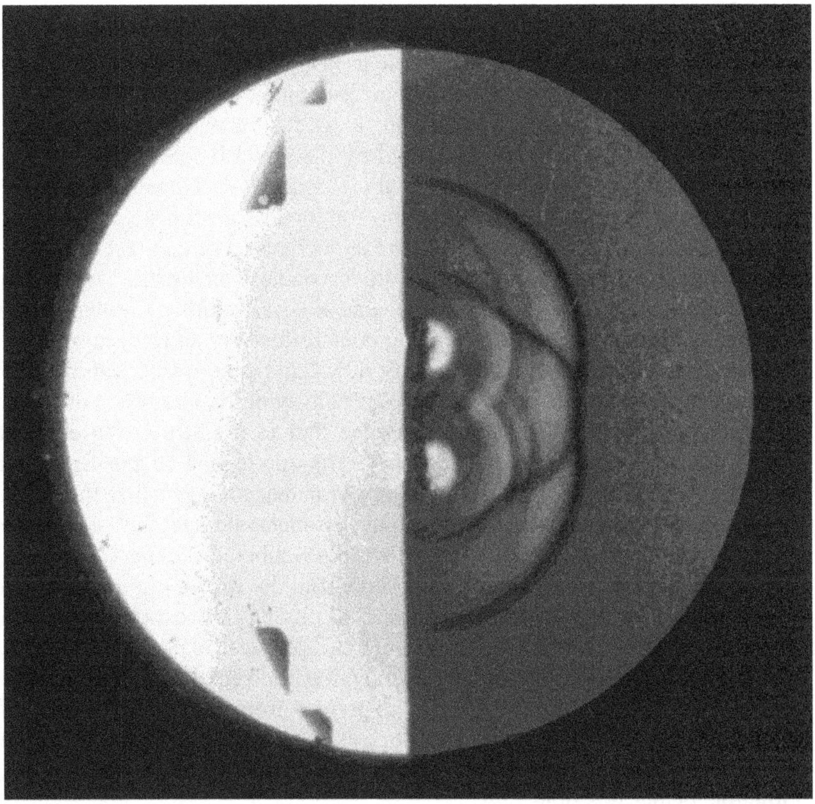

Figure 1: Double acoustic wave interference, Ernst Mach, circa 1892, (Photo Deutsches Museum ©)

[6] Mach, *op cit,* p. 310.

Mach's proposed realm of "elements" simultaneously implicates the "perceiver" and the "perceived" from a single referential construct where both the "thing" and the discreet "ego" are simply viewed as alternating pattern-bundles composed of common blocks. Elements implicate a fundamental realm that is neither physical, nor psychological, yet bridge between the two. If we were to take the elements and arrange them one way, the "self" is denoted, pick them up and arrange them again, and the "thing" appears. According to Mach, the dualistic illusion of separateness is an inevitable byproduct of the conscious mind, where these two arrangements can never coincide in a single image. Waking experience perpetuates a picture of existence that will always takes sides, and the terms "I" and "it" are simply patterns visited at a greater frequency such that they will progressively be seen to take on the characteristics of independent entities.

The Phased-space addressed in WFS, despite being invisible to corporeal sense perceptions, is accessible to instrumental observations and by extension can be seen to possess certain discernable characteristics. Such a position towards matter enforces an extended domain of "presence" stretching beyond the windows afforded by the unaided sensorium. Phased-space (if it can be considered a space at all) might be seen as a category of manifestations that are revealed if sound (or for that matter any other wave phenomena) were to peer in on itself; and as such only becomes accessible by way of translation or transposition onto secondary mediums. This realm persists at another location within the matrix of sensations – easing away from the auditory materiality the waking body reflexively ascribes to sound.

The application of photography in Mach's experiments can be seen as such a "secondary medium" capable of conveying an altogether different materiality of sound. Mach was among the first to produce cross-sectional images of the Phased-space of acoustics. His application of the Schlieren method of photography to airborne projectiles managed to visualize the shock waves in air enveloping bullets traveling at supersonic speed. Phase is the material component that shows up in Mach's Schlieren photographs as the alternating gradient bands of tone superimposing the frozen projectile onto a backdrop haze of energetic formations. Similarly, his experiments with acoustic reflections opened glimpses into the physical expanse of acoustic interactions. In consideration of Mach's approach to sensations, these images would seem to present a significant departure from the ocular model of photography practiced by his contemporaries; namely, a modality where the camera is understood to be either a partial replication of, or an extension to, normative ocular functions.[7]

[7] For instance, one could contrast what I'm proposing as Mach's photo-sensing approach to photography (and its underlying understanding a sensation-complex spatiality), with the chrono-photographic work of his contemporary Étienne-Jules Marey (where the camera is seen to intervene in the "matter" of time by breaking durations of optic spatiality into an incrementally

Even though this model was not a pronounced concern in Mach's work, I would suggest that his photographic studies of fluid dynamics in general, and his acoustic imaging experiments in particular, should be understood in terms of tactile "sensing" and not as prosthetic extensions to "vision". There is no eye that can be situated within acoustic matter, as much as there is no horizon within an acoustic expanse. The imaging capacity produced in Schlieren techniques is a useful byproduct of an intervention photography affords at a deeper level of presences.

In Mach's images of wave phenomena, the film plane acts as a partition of sorts forced between sensation interactions plumbing traces from a more primary configuration of elements. It is in this sense that his approach to sensation might be seen to include an unstated yet explicit re-thinking of the epistemic purpose ascribed to the photographic plate. Numerous photographs, directly imaging "processes" (*e.g.,* acoustic refraction, air turbulence, electromagnetic interference, and electrical spark discharge), particularly the ones taken over the course of five years spanning 1888 – 1893, reinforce this position. In contrast to normative optic models of that period, Mach arguably approaches the photographic process as an inanimate version of "sensation taking hold". Through this approach, the photosensitive emulsion is understood to be a subtle "memory substance" that engages and preserves unfolding "sensation" complexes. On the surface of the plate, the silver halide crystals suspended in a gelatin membrane become an inorganic sensitized skin and the developing process acts to preserve a cross-sectional registration of sensed processes. Possibly the most exemplary photographs of such epidermal character are Mach's spark studies where discharging electricity is sent shooting across the surface the film plate literally burning tracks into the emulsion as it travels. The eyes participation with such imprints of matter occurs only at a later stage, brought in primarily for its deciphering capacities. Mach's relation to the Schlieren images is that of discerning a "legibility of traces". He reads the surface of the image as one would read a hieroglyphic text.

Just as Mach's photographs of the behavior of sound are no longer ocular artifacts of optical space, but rather imprints of tactile interactions, such can be understood the Phased-space of sound projected by way of WFS as a reverse Schlieren image, mobilizing the gradient patches of phase back into the thickened folds of propagating waves. Similarly to Schlieren photography that does not include the configuration of the eye until after the "image" is produced, so does WFS only introduce the listener after the phase patterns have been set in motion. The sound model of WFS excludes the idea

---

sequenced appearance). Admittedly this argument demands a more careful unpacking of Mach's photographic practice assessed in relation to other prevailing normative models of photography at that time and remains beyond the scope of this current inquiry.

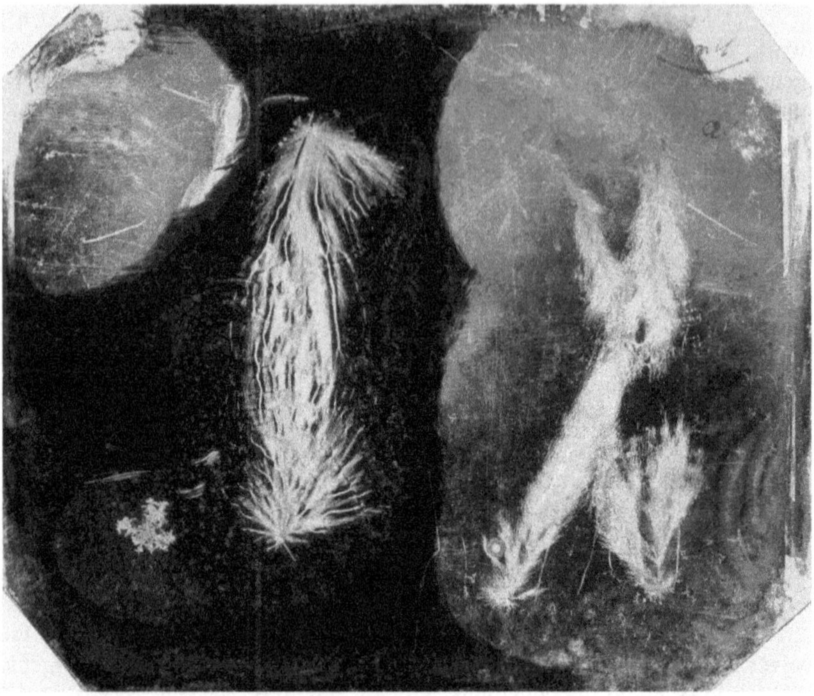

Figure 2: Spark discharge, Ernst Mach, date unknown (Photo Deutsches Museum ©)

of the listener altogether from its mode of sound production, reintroducing the ear as a non-idealized entity into a homogeneous field of listening possibilities.

One striking quality of WFS is the way in which sounds materialize in experience as distinctly spatial presences, nearly possessing dimensional attributes such as mass and length. Each position within the wave field is equally charged with acoustic formations. To hear or receive in such fields means to transfigure sonic events into acoustic intelligibility and thus transform invisible interactions into coherent localizable events. Anywhere within that field, each listener unfolds their own relational aural environment where the particular constellation of sonic events will be as unique and multiple as the number of varying points of hearing opened into the expanse of sonic interactions. To hear such configurations, means to participate in a polycentric experience, not very different from the everyday acoustic environment. As such, there is no inscribed orientation to a WFS setup, each direction is equally charged with aural potentiality. Contrary to the pronounced "frontality" that can be found in many dedicated listening spaces (particularly with regards to theatres and concert halls), in WFS there is no "front" nor "back"; furthermore, WFS has no designated location from where the sounds are intended to emerge. For the listener within this realm of acoustic interactions, every sound appears as its own unique vanishing point, balanced within its own center of gravity, and enforcing a relational framework whose definitions and boundaries are re-determined within every progressing step in the triadic relations between sound-space-listener.

Correspondences between "observer" and "observed", where audible space itself is indeterminate until the act of listening has intervened within the confines of Phased-space, underlines that there is no "absolute" sound-space from which to determine an auditory point-of-view. Phased-space is only known by way of its projection – by way of the imprint it leaves in sensing devices and by way of the audible worlds through which it communicates. Yet, Phased-space should not be thought of as detached from the framework through which we chose to extract its characteristics and behavior. No Phased-space exists "as is"; rather, there are event-observer constellations informing every unfolding position of aural spatiality.

Importantly, Ernst Mach distinguished a categorical difference between "geometric space" and what he calls "physiological space". For Mach, physiological space is the only primary domain. He argues that geometric spaces are secondary human inventions abstracted out of experiences. Other than being a useful approximation, they should be regarded as little more than conventional abbreviations serving utilitarian ends. Such intellectually derived spaces tend to perpetuate their authority through historical and cultural circumstances to the extent that their secondary nature is no longer self-evident. For example, the geometrical space of the Euclidian kind is homogeneous and unlimited, essentially an abstraction when contrasted with the space of experience. In contrast, physiological space, the space-as-lived, is heterogeneous and observer-centric. It is a highly differentiated space whose measure of detail grows with increasing proximity to the locus of observation. Furthermore, in physiological space "extension" itself is differently articulate depending on the direction of observation. In physiological space the notions of above/below, before/behind, right/left all have inherent significance, whereas they are completely superfluous to geometrical space.[8] Even when the point of observation mobilizes, and experiential space begins to approximate Euclidean space, that space will remain allied with poly-sensorial attributes, among which are colors, textures, and directionality.

From the vantage of Mach's "sensation elements", one and the same strata that lines matter and mental processes, it takes little stretch of the imagination to suggest that certain traits attributed exclusively to the

---

[8] For a description of discrepancies between "geometric space" and "physiological space", see chapter IX, "Biologic-Teleological Considerations as to Space", in Mach's *The Analysis of Sensations*.

psychological domain might maintain other, possibly inorganic, equivalences. Admittedly Mach does not pursue this path in his writings considering the dubious scientific value he attributed to such hypothetical commonalities. For example, when considering the question whether crystals have sensations, he does not reject outright such a notion, but rather states that this is a question without scientific or practical meaning.[9] One important question he is willing to partially entertain is that of "memory". In Chapter XI of his book *The Analysis of Sensation*, entitled "Sensation, Memory and Association", Mach draws correspondences between faculties of consciousness and characteristics of matter with regards to their common "retaining" capacities:

> In reality every psychical process leaves indelible traces behind, just as every physical process does. In both spheres there are irreversible processes: entropy increases, or the bond of a friendship that has been broken, and then renewed, is felt.[10]

Mach argues that the manner in which matter reveals the strain of time is apparent in such examples as geological erosion where the surface of the landscape records the passage of climatic transformations, or a wire that "remembers" the accumulated postures from its previous coilings. Yet memory for Mach is somewhat more complex in structure than the mere retaining capacities of past events. Memory involves certain instances from the past re-surfacing as a field of influence capable of directing processes in the present.[11] Possibly the most vivid example of an inorganic memory cited is that of the phonograph, though Mach adds that the phonograph requires an external force to "recall" its embedded memory whereas human memories have the capacity to play themselves automatically.

The Phased-space of acoustics may similarly be considered in terms of its "memorizing" capacities. Within Phased-space every sound is already a trace imprinted upon a flux of vibrations. In contrast to the example of erosion, the domain of phase supports memories impressed in past occurrences that have real bearing upon the outcome of present conditions. With respect to a given crest in an expanse of propagating waves, the polarity and potency that specific wave will seem to possess at any given moment in time depends both on its own energetic state, as well as on the precise choreography of encounters imposed by proximal wave events. If a crest meets another crest, the wave's energy at that point in time will seem to momentarily multiply. Conversely, if that crest meets a trough – that same wave will instantaneously vanish, only resurface moments later, after the undertow has subsided.

---

[9] Mach, *op cit,* p. 244.
[10] *ibid,* p. 237.
[11] *ibid,* p. 237.

In this sense, the characteristics of acoustic substances, themselves outcomes of sonic eventfulness, bear influence on encounters in the present tense. This form of memory would seem to be internal to the materiality of vibration itself, and our own perception of such acoustic sensations would then be yet another appearance of memory enfolding the former, thickening temporalities into spatialized durations.

Figure 3: Wave Field Synthesis acoustic rendering setup, 192 loudspeaker array (Author's design)

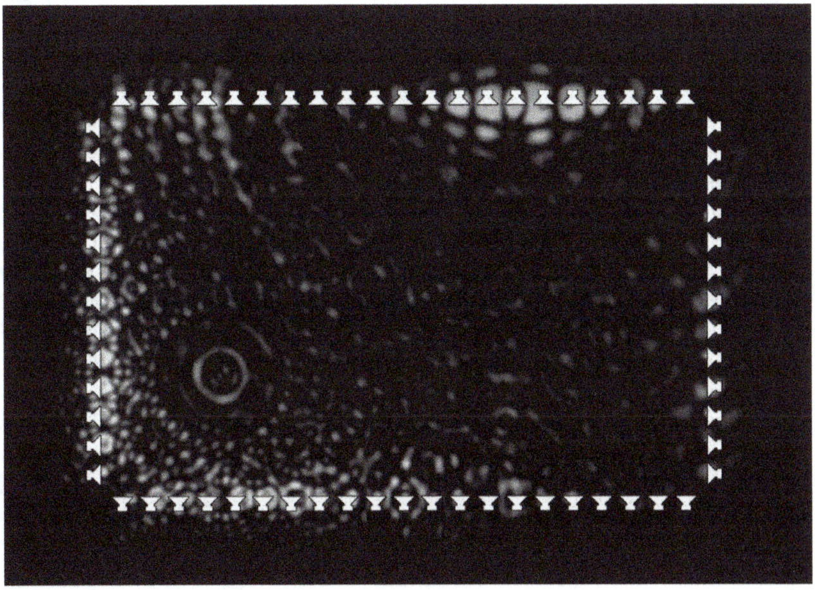

Figure 4: Phased-space sectional study, instantaneous pressure-front analysis in air of two vocal sound samples rendered via a frame array of wave field synthesis (Sound Field Plotting algorithm and image synthesis by Miguel Nigrão)

Perhaps under these particular circumstances, where memory is multiplied by memory, we can begin to explore the potency of such sonic temporalities – in the tension that arises between a memory patterning of acoustic interactions and the memory space of embodies sonic experiences. In addition, in such doublings, Mach's phonograph of inanimate memory actually begins to play and to take hold.

In consideration of such terms, WFS could be seen to set forth an expanded memory plane in search of listeners, perpetuating a domain within which every act of listening sets off a chain of delays, where perceptions enfold propagating times and spaces begin to multiply. The result is an expanded realm of potentiality for audible spaces that fade in and out of perceptual tangibility. For the listener in WFS sounds may appear to be inside, or external to, the physical location of the transducers. In certain instances, sound may even seem to extend beyond the visual confines of the performance hall itself. By way of rendering vibrations, resonating instances materializes in the absence of the very objects that supposedly produced those sounds in the first place. In other words, there is no visual confirmation for soundings rendered "present". This phantom quality is an indication of a much more mundane paradox brought to the fore in WFS: the coincidence of a listening position with a localized acoustic event is simultaneously the instance from which one derives the spatialized location of that same event. Seemingly, "to hear" means to simultaneously occupy two positions – the one on the surface of the ear, the other cast back to an imagined point of emanation. For the situated listener, the sound is at once "here" and "out there". Despite the fact that hearing is an explicitly tactile sense perception, perception will always trace that sound back to the estimated source, rarely paying attention to the proximity of vibrations. Audition, like vision, creates the impression of apprehension at a distance.

In WFS, by eliminating the visual cross-reference, one is left to grapple with the phantasmal appearances of sounds rendered tangible, albeit from the position of physical phenomenon the bind between a resonance and a location was never there in the first place. Sounds propagate by way of expansions in the cusp of events, projecting outwards in alternating patterns of residual timings. In other words, it is not the vibrating surface of objects that convey a sonic presence, but rather the *intervening distances* within which all the various vibrations combine to form a continuous cloud of eventfulness. Within this tangled matrix, the entire spatialized coherence is bundled up in an undivided shimmering mass, where each singular "boundary" is enfolded within a multitude of other singularities. The uncanny aural ability to unravel this continuity into recognizable and discreetly localized "events" happens effortlessly in every moment of listening. The tactile faculty of the ear remains for the most part shrouded beneath the listeners cognitive "space-ing" operations.

Considering frequencies in terms of their physical sizes draws attention to the materiality of waves that sheath the body with an endodermis of fluctuating eventfulness. Hearing is more than just a matter restricted to the inner ear labyrinth. Sounds are in fact sensations that travel along bones, brush up against the skin and form resonances in the inner cavities of the body. Sound's capacity to be at once immersive, intimate, and visceral; whilst at the same time being recognizable, situated, and mundane, has no readily accessible logic. It simply is, and by being such, causes audible presences to flickers in and out of synch with the other tangible spaces and things cluttering our palpable worlds. The particular materiality of sound, as we are expressing it in our contemporary auditory practices, is at once spatially differentiated and inaccessibly abstract.

It is not so much the corporealizing of Phased-space that surfaces in WFS (as that would involve surpassing the reflexive functions of hearing), but the superimposition of such understandings of the space-of-sounds back into our audible worlds. The experience of sound gets torqued by the knowledge of sounds alternate modes of existence. Sound as a phenomena remains the same, but our relation to it, the aspects we subconsciously accumulate and subsequently internalize, gives the shape and materiality that sound is seen to possess. Consequently, the spaces that sound create, and the materiality it seems to occupy, will also be in a perpetual state of reconfiguration. By the very nature of this ongoing process, hearing itself is transformed.

Figure 5: Projectile shock wave, experiment setup, Ernst Mach, date unknown (Photo Deutsches Museum ©)

In conclusion, every epoch, with its particular sonic preoccupations and acoustic materializations, inevitably hatches its own definition of a "listener". Consequently, the extents of "hearing" will be determined by the myriad of practices, techniques, and epistemologies condensing the various aural fields in any given historic and cultural context. Within our contemporary practices it would seem that the further we materialize the Phased-space of vibrational eventfulness, by way of electro-acoustic techniques as well as by way of alternative modes of vision – the more this remote aspect of sound will be seen to intrude upon the normative soundscape. As vibration itself is seen to grow skin, so will the stable confines that delineated the clear "edges" of things begin to melt back onto the flux of peripheral forces. And the faculties of perception, like a television station that has slipped out of reception, gets challenged by unfamiliar signals, interfering ever so slightly with the mundane shape of "things".

The Evidence of Film and the Presence of the World: Jean-Luc Nancy's Cinematic Ontology*

Josef Früchtl

Following Deleuze's theory, the emergence of the "time-image" following the Second World War led to (our relationship to) the world disappearing from film. Following Jean-Luc Nancy, the Iranian film-maker Abbas Kiarostami is a privileged witness to a cinema that liberates itself from a "postmodern" obsession with the disappearance of the world. Nancy reveals three foci of cinema: the look, the movement and the real (the world). The "alethic" (Heidegger) look is "mobilized" by cinema; it forces us to remain in contemplative and mental motion. In experiences of evidence the cinema presents us "the real" (Lacan). In Früchtl's article, he argues that cinema is not merely the ontological celebration of presence, but the aesthetic celebration of a tension generating and suspended difference: between presence and its re-presentation. The evidence of cinema thus is the mediated result of an interactive, even playful relationship of our dimensions of experience. What opens is a space full of possible meanings, a textual web which only takes shape with time and with the help of the subject providing meaning.

The Post-Postmodern Film

Let us begin with a brief reminder of Deleuze. Following his theory, the emergence of the "time-image" following the Second World War led to (our relationship to) the world disappearing from film. In this "post-classical" situation, film is apparently left with just two options: either it can bemoan as a loss, as an absence of meaning, the disappearance of the world and its classical film in which everything centred around stability of meaning; or, it can make this disappearance the object of an endless game. These three (pre- and post-War) historical film variations are often termed "classical", "modern" and "postmodern", but it is probably more accurate, as well as more helpful, to relinquish the modern-postmodern model, which is vague and seems to have run its course, and to call these variations "classical", "agonal" and "hybrid".[1]

Jean-Luc Nancy now appears to have added a further variation to this list. He believes that film has liberated itself from its (agonal and hybrid) obsession with the disappearance of the world by emphasising the positive side of precisely this loss of meaning and its stable ascription; yet not in a playfully hybrid, endlessly recombining sense, but in a phenomenological one. A world which no longer has any (stable) meaning, because the

---

* A longer version in the German language entitled: "*Die Evidenz des Films und die Präsez der Welt. Jean-Luc Nancys cineastische Ontologie*", will subsequently be published in: Ludger Schwarte (ed.), *Bild-Performanz. Die Kraft des Visuellen*, (München: Fink, 2010).

[1] *cf.* Früchtl, Josef, *Das unverschämte Ich: Eine Heldengeschichte der Moderne* (Frankfurt/M.: Suhrkamp, 2004), p. 19. In contrast, Laurent Kretschmar opts to retain the modern-postmodern model: "Is Cinema Renewing Itself?", *Film-Philosophy*, 6, no. 15 (July 2002).

relationship between it and its subject is no longer guaranteed, is meaningless in two senses of the word: it *no longer* has (a stable) meaning and it has no (stable) meaning *yet*. It is therefore also an entity which still has to find its (stable) meaning, still has to invent it. A world without (any) meaning would be a world in itself. It would be a *tabula rasa*, an empty page which offers itself (to writing, to ascription of meaning), a space devoid (of meaning) which opens up. And a world without a *stable* meaning in a sense restores a world in itself, namely in the sense that in the end it has no meaning at all; where nothing counts, everything is possible. This is unequivocally demonstrated, in Nancy's opinion, by the Iranian film-maker Abbas Kiarostami. Kiarostami is "a privileged witness to all this, seeing that cinema renews itself, that is to say it comes close again to what it is and yet always brings it back into play".[2] This means that cinema has an essence which Kiarostami reveals afresh, albeit not without simultaneously –and hear we hear Hegel stir– putting it at stake (*remettre en jeu*). The central concept which is to underpin this postulation is already indicated in the title of his book on the same topic. Nancy wishes to demonstrate the *Evidence of Film*.

Like the meaning of any philosophical term, the meaning of "evidence" depends on the epistemological positions assumed. The position adopted by Nancy is first and foremost clearly a definition of evidence as an objective form of truth *finding* and not merely a subjective form of truth *acknowledgement*. What is evident is the "self-revelation" of facts, and not just the "sight" of them or an "insight" into them. Evidence is one of the criteria of truth, not just one of the pragmatic constituents of argumentation. In all cases, however, (just as with the concept of intuition) its opposite number is a discursive, conceptual, methodically ordered cognition.[3] Nancy finds a bewitching metaphor for this: "Evidence always comprises a blind spot within its very obviousness…The blind spot does not deprive the eye of its sight; on the contrary, it makes an opening for a gaze and it *presses* upon it to look".[4] Evidence renders something visible, recognisable, opens up an insight, permits recognition of a truth, but not of the manner in which one arrives at this truth. The method is its blind spot, which cannot be seen because it cannot be rendered the object of conceptual determination. In the language of argumentation theory, a piece of evidence can be performatively generated but not methodically proven.

---

[2] Nancy, Jean-Luc, *L'Evidence du film/The Evidence of Film* (Bruxelles: Yves Gevaert, 2001), p. 13.

[3] *cf.* Mittelstraß, Jürgen, "Evidenz", in *Enzyklopädie Philosophie und Wissenschaftstheorie*, Vol. 1: A-G (Mannheim/Wien/Zürich: Metzler, 1980), p. 609.

[4] Nancy, *The Evidence of Film, op cit*, pp. 13 & 19.

The Look, the Movement, the Real

Thinking in metaphors and gliding along chains of significants, is a style well practised in modern French philosophy. Nancy is no exception. His composition is an essay about film which is allowed to drift in a circular current of association, revealing three foci: the look, the movement and the real (world).

The look, at least for those who think in French (*le regard*), is a particular way of looking at, or regarding the world.[5] Films, which cultivate this (out)look, "are here as eye openers, [are]...the budding and opening of a look in the middle of ordinary turbulence".[6] With this celebration of the look and the gaze, Nancy is not alone among 20th century French philosophers, but he is somewhat on the periphery. Probably with the exception of Merleau-Ponty, there are few within his tradition who are not at least ambivalent in their attitude towards the postulation that seeing is a privileged mode of cognition.[7] Generally speaking, his teachers and colleagues are united in their blackening of occidental ocular centrism.[8] Yet Nancy's book on the *Evidence of Film* comes across as completely unimpressed by this disparagement. This raises the question, of course, as to whether his stance is well-founded.

For his book on film, this question cannot be answered in the affirmative without at least some reproach. Nancy favours etymological reasons for his stance. Returning to French, the look (*le regard*) is linked to "guarding" (*la garde*) and to "looking after" (*prise en garde*). The inspiration here in both form and content is Heidegger who, alongside Derrida and Bataille, has had the largest influence on Nancy. As Heidegger wrote in his *Letter on Humanism*, the human being has "to guard" the truth of Being. The word "respect", Nancy continues, is also linked to the look, this time in Latin; *respectus* means "looking back", "rear view". And yet, as Nancy also adds, it is not the look in itself, but the "rightful look" which entails respect for the observed real, and this look is in turn distinguished as "openly attending" (*ouverture*), as an openness for the observed, which knows no distinction between defining (active) and being defined (passive).[9]

Nancy thus puts the case vociferously for a look which, in Heideggerian terms, is "*aletheic*". Heidegger himself, of course, not only famously criticized the occidental forgetting of time, the translation from *theoria* to *contemplatio*, the repression of hearing and binding (to being); he was also the philosopher of "circumspection" (*Umsicht*) and "clearing"

[5] *ibid*, pp. 15 & 39.
[6] *ibid*, pp. 17 & 23.
[7] Following Gary Shapiro, the perspective of ambivalence would be the right one, *cf.* Shapiro, Gary, *Archeologies of Vision. Foucault and Nietzsche on Seeing and Saying* (Chicago: The University of Chicago Press 2003), p. 6.
[8] *cf.* Jay, Martin, *Downcast Eyes: The Denigration of Vision in Twentieth-Century French Thought* (Berkeley, CA: University of California Press, 1993).
[9] Nancy, *The Evidence of Film*, *op cit*, p. 39.

(*Lichtung*), of a seeing which permits encounters and which at the same time unconceals and conceals.[10] This second, ontological meaning of looking is pushed to the foreground by Nancy almost ceremonially, without any further justification. Those wishing to know more need to consult books such as *Le sens du monde* and *Être singulier pluriel*, in which Nancy explains the meaning of the concept of the world, also the political-global world, amongst other things by referring to Heidegger's "there is" (*es gibt*) analysis, of being as a giving entity, (and thus to what critics of philosophical language soberly refer to as an "existence operator").

A second question remains, and this one cannot be answered by resorting to another book, namely that of why film *alone* celebrates the look and not other art forms, in particular painting or photography. Nancy insists that film is the art form of the look and that it can only be distinguished from other art forms via this characteristic.[11] His closer description shifts the emphasis to another characteristic of film, however, that of *movement*.

In Kiarostami's work, Nancy certainly has an excellent example of cinema which exudes calm, demands attention and grants its subject matter both space and time, thereby also affording it an opportunity to recede. This is cinema which even a contemplative idler like Peter Handke can really appreciate.[12] Nevertheless, in an age in which art has become manifold, in which video, performance art, body art, installation art, etc., have joined the show, the arts themselves also sustain an "inner multiplicity". This is true of "some" art forms at least, certainly of film, consisting as it does of images, music, speech and movement.[13] For Nancy, movement, the kinetic in the cinematic, is the being of cinema ("*l'être du cinema"*). Of course, for an ontologist, especially for a contemporary of Deleuze, whom Nancy quotes in this context, movement is not taken to be the object of film, something it "represents", but as something it *presents*, that it itself, ontologically speaking, *is*. "Being", according to Nancy's ontology of difference, "is not something; it is that something goes on", it is, with its moments and events "…that it continues…that it discontinues continuously. Like the images of the film".[14] Movement is "presence, which is really *present*", which means, to stay true to the phenomenological-ontological comprehension, "coming before", coming before any conceptual definition; it is something which is not given as defined, but first and foremost as something awaiting definition. It is an "opening", which is clearly not (only) meant to refer to space, but

---

[10] *cf.* Jay, *Downcast Eyes*, *op cit*, p. 269.

[11] Nancy, *The Evidence of Film*, *op cit*, p. 19.

[12] *ibid*, p. 39; *cf.* Handke, Peter, "*Die Geschichte von Hossein und Tahereh. Eine Annäherung an den iranischen Cineasten Abbas Kiarostami*", in Handke, Peter (ed.), *Mündliches und Schriftliches: Zu Büchern, Bildern und Filmen 1992-2002* (Frankfurt/M.: Suhrkamp, 2002) pp. 66.

[13] Nancy, *The Evidence of Film*, *op cit*, p. 23.

[14] *ibid*, p. 61.

(also and especially) to time, an *ouverture*.[15] What opens is a space full of possible meanings, a textual web which only takes shape with time and with the help of the subject providing meaning.

In the case of film and cinema, with its peephole atmosphere, the look is therefore an opening because it "mobilises" something,[16] it makes something move, in a multitude of senses: the sequence of images (corresponding to the mechanical sequence created by the equipment); the changing sequence of views (dictated by different camera shots); the concomitant thoughts and emotions. Being "carried away", "driven, lifted away" (*emportement*) is thus a necessary effect of cinema.[17] In this sense, for Nancy, cinema has to be comprehended as the art form of the look, as the art of mobilising, as the mobilisation of the look (*mobilisation du regard*).[18] And in this sense one can agree with him. No other art form mobilizes the look as much as film does; no other forces us so hard to keep changing our direction of gaze and to remain in contemplative (and mental) motion.

Due to the dimension of time, Nancy, in line with the entire Romantic tradition since the nineteenth century, defers the recognisability of presence issue, generally explained using evidence, to the specifically time-related mode of *suddenness*. For him, too, the epistemology of presence is connected to the dimension of the moment. *Intensity* as a mode of perception and *immediacy* as a mode of cognition ultimately correspond to this time mode. Evidence is not "what makes sense", but "what is striking" (*ce qui frappe*), and this "strike" (*le coup*) "gives a chance and an opportunity to meaning".[19]

A sense which cannot be mastered is also what is meant by the third central term --namely what is *real*. Real is what has consistency (*consistance*) and resistance (*résistance*).[20] "Consistency" is not meant here, of course, in the logical sense, but once again in the literal (Latin) sense: real is what opposes the theoretical-conceptual and pragmatic-changing intentions of subjects as a restive entity. In this interpretation it belongs to the basic stock of occidental philosophy, most recently influenced by Lacan. Like Kant's thing-in-itself, what is real is "even more puzzling than God", or more precisely that God "to whom we can ascribe particular qualities".[21] To this extent alone, Nancy's ontology is also coloured by theology and metaphysics, and increasingly so the more he addresses a different conception of God, so to speak, and poetically revolves around his objectless object of philosophical

[15] *ibid*, p. 29.
[16] *ibid*, p. 17.
[17] *ibid*, p. 51.
[18] *ibid*, p. 27.
[19] *ibid*, p. 34.
[20] *ibid*, pp. 15-16, and p. 32.
[21] Eagleton, Terry, *Ästhetik: Die Geschichte ihrer Ideologie*, (Stuttgart/Weimar: Metzler, 1994), p. 125 (English original. 1990, p. 120).

desire. Even Nancy pays homage to the negative-theological fundamental figure of recent French philosophy who is in love with paradoxes, even he skilfully plays the whole range of *amour fou*, which cannot live *with* the metaphysics of the absolute, but cannot live *without* it either. But he is far less happy than Derrida[22] (or Adorno, in the German context), to be a prisoner of this system of thought. Like Deleuze and Badiou, he searches for the positive, for ways out, for loopholes and gaps, and *his* names for these are "evidence", "presence", "world".

Some Unclarified Points

In order for the positive to fulfil its function, it requires at the very least a sufficient definition. In this context, I see in Nancy's deliberations several unclarified points. First of all, to be fair, Nancy should point out that determining the essence of film as mobilisation of the look was not his own discovery. This honour goes to Erwin Panofsky, who in his essay "Style and Medium in the Motion Pictures" names as the two specific achievements of film the "dynamisation of space" and the corresponding "spatialisation of time". Accordingly, the fixed place that a spectator assumes in his cinema seat should not hide the fact that, as an *aesthetic* spectator, he is "constantly moving", "his eye identifying with the lens of the camera". Unlike on the theatre stage, "not only are bodies moving within space, but space itself is moving, approaching, receding, turning, dissolving and reforming."[23] The mobility of the look is one of space. In the context of film, "seeing" particularly means seeing and experiencing a permanently changing space which is bound to time. Film provides a mobile, picturesque experience of space, a mobile experience of space through pictures, the experience of a virtual mobile space.[24]

Secondly, there is an astonishing one-sidedness in Nancy's definition of the essence of film. He comprehends "mobilisation" of the look only in its literal sense. It is telling that he ignores its *military* sense, even though this addresses the aspect undeniably central to film theory that, with the exception of music, film proceeds more dictatorially than any other art form. Film imposes a look upon its recipients (just as music imposes a sequence of sounds). Its *alethic*, "being-presenting", and thus "truth-unconcealing" dimension is not to be separated from its concealing dimension, not only in the ontological-philosophical, "dialectic" sense made

---

[22] Derrida, of course, rejects the objection that his thinking of , '*la différance*' is negative theological. See his famous article "Différance", in *Margins of Philosophy*, translation by Alan Bass, (Chicago: The University of Chicago Press, 1982),p. 6.

[23] Panofsky, Erwin, "*Stil und Medium im Film*" in *Stil und Medium im Film & Die ideologischen Vorläufer des Rolls-Royce-Kühlers* (Frankfurt/M.: Suhrkamp, 1999), p. 25.

[24] *cf.* Martin Seel, "Thirteen Statements on the Picture", in *Aesthetics of Appearing*, translation by John Farrell (Stanford, CA: Stanford University Press, 2005), p. 182, with references not only to Panofsky, but also to Noël Carroll.

famous by Heidegger, but also in the crudely manipulative and thus political sense.

This denial of the ambiguity behind the term mobilisation is in line with a third objection, or at least a third unclarified point. According to Nancy, in evidence, "in the image's strength", "a shelter from the image's ability to betray" also lies.[25] Nancy, unfortunately, leaves this as a peripheral comment, when on this point we would like to know so much more. The "ability of the image to betray" consists in projecting something as being which is not. And the question is how evidence is to provide "shelter" from that. For it can only lead to the experience *that* something is, not *what* it is. And yet, in cases of betrayal, the *what*, the false or deceptive defining of an (undefined) given, is usually crucial. But even remaining at the level of pure presence, the problem of distinguishability is still there. Even the experience *that* something (undefined) is there can be doubtful, an experience expressed in the question: "Is there anything there (at all)?" or in the statement: "There seems to be something here". The "ability of the image to betray" then consists in the fact that it arouses an *appearance* of presence. The age-old question in this case is therefore whether and how to control appeals to evidence or, put another way: whether and how evidence can be distinguished from apparent evidence. Nancy does not even acknowledge this question. His only fleeting contact with it is when he, as already quoted, speaks of the "correct" look, necessary in order to testify respect for the real.

Fourthly, the relationship between presence and representation, between experience of evidence and articulation of sense, remains unclarified. Nancy's ontology, influenced by Heidegger, leads us to expect a precedence of presence and evidence over the "descended" and secondary representation and sense. In fact, Nancy at one stage describes the relationship as one of *tension* and *suspension*, which can be interpreted as a relationship in which both sides have a mutual negative reference which overrides their one-sidedness. Put another way: a relationship in which their mutual reference might be negative, but it is also necessary, and through which they generate a tension (in both a structural and an emotional sense). Cinema "stretches and hangs between" (*est tendu et suspendu*) the world of presence and that of representation; it is (literally) stretched like the "screen" which Nancy (figuratively) observes to be a "sensitive membrane", a skin mediating between two worlds, the presenting world and the representing world, doing so in such a way as to leave both in the balance.[26] Seen in this way, cinema is not merely the *ontological celebration of presence*, but the *aesthetic celebration of a tension-generating and suspended difference* - between presence and its re-presentation. This would, of course, also have *epistemological* consequences, the most important being that a pure, pre-

[25] Nancy, *The Evidence of Film, op cit,* p. 35.
[26] *ibid*, p. 57.

conceptual, sense-free experience of evidence would have to be characterized as an empty structure, or more precisely that we might have to *presuppose* the presence of (objects and events within) the world, but that we cannot keep *experience* of them/it sense-free.

The last unclarified point concerns the status of the evidence experience. Nancy switches back and forth between an *ontological* and an *aesthetic* conception. Initially it is clear that evidence is aimed at proving the existence and presence of phenomena. But in his essay on film, Nancy explicates and develops his ontological understanding within an aesthetic context, rendering it unclear how these two levels relate to each other. Like every art form, film is a "reconfiguration of experience and therefore of the world".[27] Experiencing the world (having it presented) through film is a different matter than experiencing it through another art form or from the outside, beyond art altogether. But with Nancy, the dimensions of experience flow into one another. When, for example, he describes "the constants of the landscapes" in Kiarostami's films as "alternating rhythmically in wide shots which sometimes stand still and sometimes seem to be fleeing from a fast car", for him they are not "accompanying an action", but "wide presences" (*larges présences*).[28]

This levelling out of the differences inevitably leads, however, to an objection based on another conception of aesthetic experience inspired by Kant, Schiller and Dewey. Evidence, the power of conviction and persuasion within an aesthetic experience, on no account has to be due to an immediacy or a directness, but on the contrary can be due precisely to an interactive, even playful relationship. Accordingly, evidence is *feeling*, which is based on a dynamic of mediation, a movement which swings between our various dimensions of experience. Aesthetic evidence is a holistic experience which is not *pre*- but *trans*discursive. This conception clearly needs to draw upon a number of theoreticians, as indicated above, in order to be convincing in its turn. And a detailed discussion would clearly have to commence at this point.[29] In the present context, however, my insistence upon Nancy's conception of evidence being subjected at least to a *differentiation* should suffice.

Finally, I would like to raise a minor doubt concerning the "evidence machine" named cinema or film. It is doubtful whether film *per se* presents the world and can rely on this ontological power. I would like to substantiate this by turning to *Un long dimanche de fiançaille* by Jean-Pierre Jeunet. This film is about a long-awaited reunion with a much-loved person, in other words an extremely affective experience of presence. When the

---

[27] *ibid*, p. 21.
[28] *ibid*, p. 57.
[29] I presented this conception under the title "Integrative Aesthetics" in the first chapter of my book *Ästhetische Erfahrung und moralisches Urteil* (Frankfurt/M.: Suhrkamp, 1996).

young wife, standing in a sun-drenched garden, eventually sees her husband again, whom she has traced down using detective work following the First World War, and whom she never once believed dead, the camera retreats almost hastily. Off-screen the voice of the narrator keeps repeating the same sentence: "*Elle le regard, elle le regard ...*". This look belongs to *her* alone, not to *us*, the viewers. The camera has no images to convey this look, *le regard*, in this emphatic, paradisiacally enchanting sense, this wordless and endless gazing. It is not suited to this type of presence, the presence of a happy ending. Happiness, or the overwhelming experience of presence, is not describable and not depictable. In this context, too, the very least that can be insisted upon is *differentiation*: "presence" is not always "presence"; it is revealed in different forms, not all of which are suited to pictures and cinema.

Societies of Control and Chrono-Topologies

M. Christine Boyer

In this chapter, Boyer argues that cybernetics, systems theory, computer programming - in short, the information mechanisms implementing "societies of control", have everything to do with time, as well as space, as Foucault outlined. Time is involved in a signal relayed, a message sent out from one post to another where it is received, decoded, interpreted. The arrow of time is always involved in control theories of communication – there is noise in the transmission, elements received and elements not received. Furthermore, in a cybernetic control system, learning is a matter of feeding back information from the environment to the system in order to regulate its dysfunction, or entropy; *i.e.*, disorganization or chaotic behavior of the system. Both feedback and entropy deal with the arrow of time – one reflexive and corrective, the other disintegrative, a matter of slowing down and disorganizing matter. Precisely these imagined connections between art and time, architecture and communication that worried theorists in the 1950's and 1960's, for might these shifts towards an information control mean that the object of architecture was dematerializing before their eyes? Boyer examines several contemporary thinkers in order to draw an analogy with chrono-topologies, or a complex of different spaces and separate times.

"Societies of Control", as Michel Foucault described them, belong to the modern space of emplacement.[1] We do not live in a spatial void in which things and individuals can be placed, but instead we reside inside an ensemble of relations that define emplacement. Foucault formally depicted such ensembles of control as relations between points in a network, elements in a series, and nodes on a tree - all icons of information processing deployed in the 1950's and 1960's. Foucault used for an example of a network, the storage of information in the memory of a computer; for a series, he deployed the circulation of discrete elements with random outputs such as automobiles on a highway; and for a tree, he examined the coded elements of an archival structure assembled according to various categorization, classification and distribution schemes. All these icons were ensembles of control, controlling how knowledge was gathered, shaped and disseminated. Highways or computers, for example, did not confine people in places - but they did multiply means of control. One could travel freely about on a roadway while being perfectly controlled by speed limits and police surveillance; one could generate information from within highly controlled access points such as libraries or museums.

Gilles Deleuze, as the major interpreter of Foucault's "Societies of Control", maintained these societies were distinct from, yet overlapped with,

[1] Foucault, Michel, "Of 'Other' Spaces", *Diacritics* 16, 1 (Spring, 1986):22-27; Michel Foucault "Other Spaces: The Principles of Heterotopias", *Lotus International* 48/49 (1981): 9-17.

the closed "Disciplinary Societies" Foucault's work normally deciphered.[2] The latter deployed their grid of power/knowledge via places of confinement in prisons, schools, or hospitals. To the contrary, "Societies of Control" no longer relied on disciplinary procedures; they had no need to confine bodies in place, but opened onto continuous flows of information, bodies, and objects.

Foucault took the label "Societies of Control" from William Burroughs whom he admired, or so Deleuze reported.[3] Burroughs was obsessed in the 1950's with how language, education, and conceptual structures conditioned beliefs, determined reality, created a type of thought-control more pernicious than disciplinary laws and regulations. Foucault agreed with Burroughs that the future would be more controlled than disciplined.[4] The future would produce new visibilities, new utterances, new power relations, new forms of subjectivity requiring new forms of resistance. Eventually, "Societies of Control" would engender new discursive and imaginary regimes of information.

What does this shift from "Disciplinary Societies" to "Societies of Control" mean for architecture, and even more pertinent what does it mean for architectural historians? And what does this have to do with chrono-topologies? This paper will argue that cybernetics, systems theory, computer programming, in short the information mechanisms implementing "Societies of Control", have everything to do with time, as well as space, as Foucault outlined. Time is involved in a signal relayed, a message sent out from one post to another where it is received, decoded, interpreted. The arrow of time is always involved in control theories of communication – there is noise in the transmission, causing some elements to be received other elements lost or garbled. Furthermore, in a cybernetic control system, learning is a matter of feeding back information from the environment to the system in order to regulate its dysfunction, or entropy (i.e., disorganization or chaotic behavior of the system). Both feedback and entropy deal with the arrow of time – one reflexive and corrective, the other disintegrative, a matter of slowing down and disorganizing matter. How then are messages transmitted over time, how might information be packaged more effectively so it reached its goal, how might new communication media be deployed to enhance spectator learning or reception, how might behavior be controlled in new and different manner? It is these and many other imagined connections between art and the flow of time, architecture and the transmission of information control and freedom, that both worried and excited artists and historians in the 1950's and 1960's.

---

[2] Deleuze, Gilles, "Postscript on Societies of Control", *October* 59 (Winter 1992): 3-7.

[3] Deleuze, Gilles, "What is a Creative Act?", *Two Regimes of Madness*, The Revised Edition of *Texts and Interviews 1975-1995*, edited by David Lapoujade, translated by Ames Hodges and Mike Taormina (Cambridge, MA: MIT Press, 2007), pp.312-324.

[4] Deleuze, Gilles, "What is a *Dispositif?*" in *Two Regimes of Madness*, *op cit*, pp. 349-348.

Might these shifts towards information control mean that the object (of art and architecture) was dematerializing before their eyes, disintegrating into bits and pieces of an information code, but might it also open up new networks of communication, new modes of reception, new systems of knowledge and communication? The path ahead for "Societies of Control" was yet unknown, but its layers of chrono-topologies was the field to explore.

George Kubler

To draw the analogy with chrono-topologies, or a complex of different spaces and separable times, let me begin with George Kubler's very influential, oft-translated, albeit opaque book of 1962: *The Shape of Time: Remarks on the History of Things.*[5] Kubler was an art historian and specialist in Meso-American artifacts, who taught at Yale University from 1940 to 1983. His theories on multiple spaces and times offered an escape from the fixed rigidity of iconographic theories of art, those which search for the moment of discovery and its successive transformations, those linear histories which left so many topics and objects outside the discussion of art.[6] More importantly, Kubler deployed communications theory in order to describe how historical change operated, how works of art were transmitted over time. He argued that historians contribute to the manifold shape of time; their aim, moreover, is to portray time – not biological time which is a continuous unbroken line stretching from birth to death, but historical time with all of its interruptions, variations, and indeterminacies. Historians enable the "shape of time" to emerge, they craft credible messages from weak signals, they project beginnings and endings into the present from hazy planes, they link together a broken but much repaired chain of objects.

Outlining the contours of a type of "chrono-topology", although clearly not deploying the word, Kubler noted that:

> Historical knowledge consists of transmissions in which the sender, the signal, and the receiver all are variable elements affecting the stability of the message. Since the receiver of a signal becomes its sender in the normal course of historical transmission (*e.g.*, the discoverer of a document usually is its editor), we may treat receivers and senders together under the heading of relays. Each relay is the occasion of some deformation in the original signal. Certain details seem insignificant and they are dropped in the relay; others have an importance conferred by their relationships to events occurring in the moment of the relay, and so they are

[5] Kubler, George, *The Shape of Time: Remarks on the History of Things* (New Haven: Yale University Press, 1962).

[6] For Kubler's influence on Robert Smithson's work see, for example, Lee, Pamela, *Chronophobias: On Time in the Art of the 1960's* (Cambridge, MA: MIT Press, 2004).

> exaggerated. One relay may wish for reasons of temperament to stress the traditional aspects of the signal; another will emphasize their novelty. Even the historian subjects his evidence to these strains, although he strives to recover the pristine signal.[7]

Kubler was even more specific about how works of art change over time: "In the first place, the historian's idea of change is related to the linguist's idea of "drift", exemplified by the progressive separation that widens between cognate languages. This "drift" produced by cumulative changes in the articulation of sounds, can be related in turn to the interferences that distort any audible communication. The telephone engineer calls such interferences "noise." "Drift," "noise," and change are related by the presence of interferences preventing the complete repetition of an earlier set of conditions.[8] Hence broken or missing links in the chain of transmission are inevitable.

Thus as works of art communicate over time, their signals break up into unintended messages, some surviving others not. The work of art for Kubler is both a material object and a message, or information. Again Kubler relies on communications theory declaring that: "…works of art resemble a system of symbolic communication which must be free from excessive 'noise' in the many copies upon which communication depends, in order to ensure some fidelity…".[9] Indeed, he goes on to argue, "…a signal trait of our time is an ambivalence in everything touching upon change. Our whole cultural tradition favors the values of permanence, yet the conditions of present existence require an acceptance of continual change."[10]

Clearly, science and technology were transforming the writing of history within "Societies of Control" as they emerged after WWII. Kubler was unconsciously deploying Foucault's series and sequences, networks and trees, interjecting mathematical notions into the description of historical change. He often utilized mathematical concepts and graphs: "[t]o speak of sequences or series, that is, of specified needs and their successive stages of satisfaction, is to mark a variety of durations."[11] But to speak of duration means also to discuss earlier and later moments in a series, and to note that earlier precedes later, and hence to link the "systematic age" to the position an item holds in a formal series of duration. Kubler goes further in his mathematics of transmission, noting that in every historical moment there simultaneously exists old and new series, but often not in synchronization with each other. He defined a time series of disparate events, occurring across vast distances yet in simultaneous times.

---

[7] Kubler, *op cit,* pp. 21 – 22.
[8] *ibid,* p. 60.
[9] *ibid,* p. 61.
[10] *ibid,* p. 62.
[11] *ibid,* p. 53.

> ...We cannot clearly descry the contours of the great currents of our own time: we are too much inside the streams of contemporary happening to chart their flow and volume. We are confronted with inner and outer historical surfaces. Of these only the outer surfaces of the completed past are accessible to historical knowledge.[12]

Kubler was alluding to the fact that he could not clearly outline the contours of the emerging "Societies of Control", he could not actually predict what the communications theory, the shifts in science and technology, were predicting for the study of messages from the past, or the drift of information about series and duration, about what comes after the past and the present. The outline of the future remained uncertain and indeterminate.

In 1965, Kubler was asked "what can historians do for architects?"[13] He replied with remarking upon a clear mistake: the banishment since 1935 of historical studies from American schools of architecture. Modern architects wanted their work to be accepted, yet they had to wait until at the gates of perception until it slowly changed. Architects wrongly believed they could reduce the number of inbound messages by increasing the amount of what they were willing to discard. Hence they restricted the traffic at the gates of perception admitting only pure and simple forms. The role of historians, Kubler reiterated, is to discover the "manifold shapes of time", not in order to recreate and re-enact these shapes, but to compose meaning from a tradition, to communicate a pattern. The historian is also a policeman, on watch for the public's foolish and implausible interests. But the true role of the historian is to sweep up the gigantic profusion of cities and things, to chart their route, to cut back unwanted growth, to teach the architect that no art is ever finished. Every work is only a work in a sequence; each rests on prior achievements, and is only a moment in a process.

Reyner Banham

We turn next to consider the British architectural and design historian, Reyner Banham who was also influenced by the information revolution taking shape after WWII. In *Theory and Design in the First Machine Age* published in 1960 Banham wrote:

> This book was conceived and written in the late years of the Nineteen-fifties, an epoch that has variously been called the Jet Age, the Detergent Decade, the Second Industrial Revolution. Almost any label that identifies anything worth identifying in the period will draw attention to some aspect of the transformation of science and technology, for these transformations have powerfully

[12] *ibid,* p. 31.

[13] Kubler, George, "What can Historians do for Architects?" *Perspecta* 9 (1965): 299 – 302.

> affected human life, and opened up new paths of choice in the ordering of our collective destiny.[14]

The following year, Banham outlined the impact of science and technology had on the architectural historian. Not only was the future indeterminate, but the importance of architecture was minimized and the role of the historian turned into one of psychoanalyzing the repressions and omissions of the architectural profession. Banham explained:

> History is to the future as the observed results of an experiment are to the plotted graph. That is, you plot on the graph the results of which you are sure, you seek for a line, an algebraic curve, that connects them convincingly and you produce it beyond the last certain point to see where it will lead. So too with all major works of historical philosophy; they extrapolate present trends into the future condition of men....The real world is rarely as tidy as mathematics, and often has a dirty trick held behind its back --- it is only at the last minute, so to speak, that the observed results begin to suggest that aerodynamics will go wild just before the speed of sound: that the curve will suddenly develop characteristics which you could hardly have expected from earlier results.[15]

Unpredictable, out of control events, made art no longer central to existence, pushed to the margins by science and technology. There were plenty of changes to perceive in the surrounding "real world", but the artistic avant-garde may have reached a dead end. Like Kubler's "shape of time", Banham is questioning formal sequencing that enabled future predictions, that controlled the forward trajectory of thought. But art, and implicitly architecture, were now sending out weak signals from the past across a void that created a reception failure and in its wake the projectile of future events might go ballistic. Banham relates:

> So, architecture is no longer central to the business of building but has become a marginal or luxury activity in this field....architecture, as the process of creating buildings that are works of art, has become empty for some reason, the actual

---

[14] Banham, Reyner, *Theory and Design in the First Machine Age* (Oxford: The Architectural Press, 1960) ,pp. 9-10.

[15] Banham, Reyner, "The History of the Immediate Future", *JRIBA* (May 1961): 252-57. Quotation on p. 252.

> business of design has ceased to be a worthwhile or interesting occupation for the mind.[16]

This calls for a reversal of how architectural history was written, sustaining a more ironical pose, destroying linear iconic accounts, introducing new information. Were there other signals from the past that might be recognized if the dominant ones no longer emitted strong messages, no longer fascinated the mind of the historian? Implicitly, it is the kind of information that architectural historians relate that informs Banham's complaint. Instead, or so he professed:

> ...the new historiography puts the movement on the couch and asks embarrassing questions. A father-figure like Auguste Perret who appears as a scion of a noble line of classicists on one count, may appear as a structural pervert on the other; an influence like Mondrian, whom the "official" view may see as an exemplar of simplicity and purity, may be seen from the other side as a man in the grip of an almost catatonic fixation, unable to do anything but repeat the same image over and over again.[17]

We might consider Banham's position to be a forerunner of post-modernism's "self-reflective turn", or "recursive thought" – a feedback loop re-directing the trajectory of architectural thought to keep it from going wild.. He was calling into question the very sources that created the history of the pioneers of modern architecture as Nikolas Pevsner had taught, turning criticism back on itself to examine the afterimage of the modern generation and what appeared outside of its official rendering, what information it had omitted or repressed. Like Kubler, he too was aware that the chain of history had missing links. Yet Banham at the same time was promising to liberate the profession from the logic that made it function, the dogma that Pevsner outlined as the linear route of "modern architecture's" development from 1890 to 1914.

Architectural historians such as Pevsner or Siegfried Giedion – as far as Banham was concerned, created "zones of silence" based on two misconceptions: first, that there were scandals to be hidden; and second, by continuously repeating the official line that the Modern Movement was a direct continuation of Rationalism and Functionalism of the 19th century. Banham believed this insistence to be a polemical device assuming that modern architecture was still useful; *i.e.,* stressing its "honesty about function, materials, structure, respect for hygiene, economy and rationalization of

---

[16] Banham, "The History of the Immediate Future", *op cit,* p. 252.
[17] *ibid*, p. 332.

construction."[18] This device was the type of linear evolutionary history, the biological metaphor, which Kubler had argued against - one that rejected the "chrono-topological" consideration of simultaneity in both time and place, and of eruptions and discontinuities in the chain of inheritance.

By the 1950's these "zones of silence" were evident and a new generation of historians began to fill in the details that had been silenced. Banham looked at the work of Walter Gropius in early 1920's, calling it an embarrassment to the dictates of modern architecture, and thus it was subsequently repressed. His Sommerfeld House of 1921 was an expressionist log cabin, and made nonsense of Pevsner's supposed continuity between pioneers and master phases. "Was this a psychotic aberration or craven submission to the importunities of his client?"[19] There were other aberrations of Futurism, or Expressionism in the town planning projects of Bruno Taut, Mendelsohn's Einstein Tower, or the writings of Paul Scheerbart. Banham claimed architects needed all the facts about modern architecture, a clear understanding of what really happened, not a polemical discourse whose aim was to secure the future of "modern architecture" in its embattled position after WWI. "History, considered in this light is not, in Alan Colquhoun's telling phrase about Giedion, 'The question for respectable grandfathers', but a psychiatric inquiry into the springs of action, the grounds of inhibition."[20]

If the historian's assignment had been to describe the evolution of history, the generations of fathers, and whom they begat, Banham instead wanted to break this line of history and discover its ruptures and holes. He wanted to put the profession on the couch, yet "...the appointment of historians to a cure of souls, to the guardianship of the conscience, even the sanity, of the profession, places upon their shoulders a responsibility that they have not been asked to carry before."[21] The architectural historian will have to grope forward "[w]ithout the ballast of an equivalent millennial tradition, architecture will have to be consciously trimmed and steered as it proceeds, and someone will have to plot its course continually. That someone is the historian; it is not for him to give orders or indicate destinations, but his plot of the track to date must be accurate."[22]

Banham is tinkering with the "time" of history - reflecting back over how the history of modern architecture came about and rewriting that history – adding new layers to the chrono-topologies of the 20th century. Unbeknownst to him, he was following the historian's model that Kubler had

---

[18] Banham, Reyner, "History and Psychiatry", *Architectural Review* (May, 1960): 326-332. Quotation on p. 326.
[19] Banham, "History and Psychiatry", *op cit,* p. 330.
[20] *ibid*, p. 331.
[21] *ibid*, p. 332.
[22] *ibid*, p. 332.

described. He was also sensitive to the shift towards "Societies of Control" occurring in the postwar decades, in particular how this translated into new design methodologies.

When Reyner Banham wrote "Softer Hardware" in 1969, he believed the factory system was outmoded, yet it still controlled most designers' mentalities.[23] This fact bound them to the old-fashioned belief that raw materials were taken to a factory, assembled into objects, and then distributed to places of storage and consumption. This system involved the use of humans at only two points: the producer and the consumer of goods. Whatever the reason for this separation in time and place, by the end of the 19th century, very few people had any idea of how production actually operated.

Finally in the 1920's - as Banham's story relates - the machine entered the domestic interior. Designers of the twenties had as small children seen at eye-level captivating machines enter the home around 1900 such as vacuum cleaners, phonographs, the telephone, new clean electric lights. In the 1920's, therefore, they praised the Bauhaus for creating cheap and elegant lighting fixtures, for standardizing goods and bringing them into the home, although they forgot to consider the human element. They never noticed that these fixtures caused intolerable glare in domestic interiors and soon were relegated to the junk heap or secondhand shops. The gap between the intent of designers and the desire of consumers began to loom ever larger. Without stating it explicitly, Banham is referring to a missing feedback from consumer desires back to designer intent that would keep the trajectory on target, the feedback loop that guides the projection.

Banham, wanting to close this information gap, turned to examine the USA, finding there the transformations he was looking for – a post-industrial era where machines and technology were small enough to be portable, even held in the hand or able to be plugged into any socket anywhere in the house. American designers such as Ray and Charles Eames, or Nelson Lewis were at home with this new technology; they played with it, experimented with it, and turned it about. In addition, Banham found the material base of this new technology increasingly invisible. A transistor radio, for example, depended on immaterial air waves; heating equipment disappeared into the walls leaving only a thermostat visible.

These transformations – or so Banham believed - should affect the mentality of designers in Europe as well as USA. Designers must forget the factory system idea where objects are produced at a distance from the life and control of consumers. Instead, they must deploy an interactive feedback system responding to customers' desires. Perhaps it was Ford, or GM, who first displayed this new interactive system when they stopped mass producing

---

[23] Banham, Reyner, "Softer Hardware", *ARK* (Summer 1969): 11-19.

standardized automobiles, and began to design only a series of interchangeable options, allowing a customer to assemble the parts as desired. A series of ready-made components could be put together in innumerable ways requiring only the support of an information-retrieval system – or catalogue – that gave the customer information to select the right parts to assemble into a Ford motor car or a specialized hot-rod. Unique to this awareness of softer hardware was the understanding that the end product was only a means to a far greater end: not a means to travel about, but a means to catch a girlfriend, show-off in front of one's neighbors, or shock an older generation. Whatever the end goal, it was far too specialized to be met by a standardized mass produced object.

Here is the arrow of time moving from consumer's desire to designer's intention generating, or at least as Banham saw it, a post-industrial society that could break free of "the cast iron prison of the factory system".[24] Clearly this was evidence of Foucault's shift from disciplinary societies to ones of control; Banham, while not mentioning Foucault directly, notes that many critics especially David Reismann, author of *The Lonely Crowd*,[25] believed the computer and the information society would dramatically increase both the detail and the closeness with which everyone's lives were brought under surveillance and control. Many were fearful of the development of electronic control networks because their operations, or so Banham believed, were completely invisible, like the example of heating systems offered above. This invisibility did indeed have menacing implications, for Banham noted that this disappearance of social control mechanisms produced fewer visible points as targets of resistance or opposition.

Here too a shift had to be recognized. In order to be effective, these control networks had to be open to public access. There had to be points of intervention if the system was to be truly interactive between designer intentions and consumer desires. But this engendered a new threat also: anyone could disrupt this system, if they simply acquired information about the complex and cross-linked networks on which they depended. Resistance no longer was a matter of mass protest, nor physical strength, but knowing how to sever the right links in the right place at the right time. If information ceased flowing, then the entire system would fail.

Banham returned to his story about what we now call "user friendly" operations. He noted that existing networks of command and control plus techniques of information gathering by manufacturers were increasingly inadequate. They did not understand what people were doing with the products they marketed, how they were transforming their products via a mixture of software and hardware until the end assemblage fit their desires.

---

[24] Banham, "Softer Hardware", *op cit,* p. 18.
[25] David Reismann, *The Lonely Crowd* (New York: Double Day/Anchor, 1952).

Only if designers broke out of their functional mentality, only if they stopped designing a mechanism to meet an isolated function, subsequently refining that mechanism towards boringly anticipated perfection, would they understand how to integrate users into the system. The triumph of software over hardware meant that the user became an adaptive part of the system of operations, that these systems not only liberated the user from repetitive tasks, but freed them from the fixed image and fixed mono-functions of designed objects. Instead, softer hardware created "indeterminate participatory open-ended situations".[26] The world of discreet, timeless, and fixed objects was replaced increasingly with a world where objects, or at least their images, captured the flow of time in motion.

Banham's definition of the historian's and the designer's role outlines a new process, more open-ended and participatory between designers and consumers, between the architectural object and the spectator. If the architectural historians he cited were interested in a factually "correct" field of objects, then Banham was, by contrast, interested in channels of communication, a behavioral context in which change might occur indeterminately and unexpectedly. As the feedback loop draws the consumer/spectator into the process of design, a dialogue is established, ideas and information set in motion, the historical trajectory and the end result continuously at play. The cybernetic model of feedback generates a series of unexpected links, repressed associations, and an unfamiliar relationship, for this was the ground Banham explored.

Robin Evans

Turning to another example of the emerging "Societies of Control" impact on architecture and design, let us take up Robert Evan's essay "Interference: Towards Anarchitecture" also published in 1969. This article is drawn from Evans' fifth year thesis project at the Architectural Association (AA) entitled "Toward Anarchitecture: Artifact Systems with Respect to Human Freedom".[27] Evans is concerned in this article with the relation between architecture and human freedom, and the need to rethink that relationship. The article is implicitly about systems analysis, information feedback loops, and self-correcting mechanisms under "Societies of Control". The article also reads as a commentary on Banham's "Softer Hardware" article, for Evans also focuses on the contemporary design of either small and large apparatuses or "artifact systems". He lays stress, however, on the socio-political

---

[26] Banham, "Softer Hardware" *ibid*, p. 17.

[27] Evans, Robin, "Towards Anarchitecture", *Architectural Association Quarterly* 2;1 (January 1970): 58-69, also in Evans, Robin, *Translations from Drawing to Building and Other Essays* (Cambridge: MIT Press, 1997), pp.11-33. This talk draws on a paper by Joy Knoblach, "Visualizing Anarchitecture, Utopia, Cybernetics, and the Supreme Shrug", *Arc* 572 (Spring 2007).

implications of these "things". He is against architecture's desire for order and control, hence "*ana*-architecture", and for indeterminacy and openness.

In the past "things", such as the railway network, telephone system, domestic electricity were difficult to analyze, complicated by the fact that utilitarian concepts such as "function", "needs" and "movement patterns" gave the designer no information about the impact these "things" had on human behavior. Instead, designers imposed images of order, utopia, or flexibility on people's behavior; time to move beyond these impositions. Instead, Evans rather humorously banishes this "putrescent terminology" to the "Wicked Word Disposal Cylinder", and thinks all modernists' attempts to impose archi-types on everyone is both elitist and repressive.[28]

Instead, architects should focus attention on how "things" effect human actions: do they allow for choice, for freedom, or are they constraining and restrictive? But here too these are no simple concepts. Modern architects may believe they design for maximum "choice" or "freedom", and they may think that changing the world to allow such to occur is a simple matter. But this assumption is wrong in Evans' opinion because change to a system of organization paradoxically entails interference and may not be beneficial. In other words, the feedback loop of information may entail positive or negative correctives; it may generate noise in the system.

There are, for example, positive interferences that expand possible actions but do not restrict existing ones. The telephone network, for example, does not prevent actions that would have happened before its invention, yet simultaneously allows for novel actions such as instant communication. There are negative interferences, however, restricting possible actions, and offering no alternative or new actions in their wake. For example a "prison": its walls restrict certain actions but do not expand on possible actions. In addition, there are synthetic interferences combining both positive and negative interferences; they restrict the set of possible actions while adding novel possible actions. Evans uses the example of an ordinary road. A road offers more free time by cutting commuting time therefore it is a positive interference, but due to heavy traffic, children can not cross the street by themselves and have to be transported to and fro; therefore, it is a negative interference. In other words, Evans is attuned to how information is sent out over the wires, how it meets with resistances – he even utilizes the image of electronic circuitry with negative and positive feedbacks and resistances, noise, transformers, switches to transmission. He is like Kubler, aware of information theory and how it influences the manner in which architecture

---

[28] Quote by Knoblach from Evans, Robin, *Towards Anarchitecture. Artifact Systems with Respect to Human Freedom.* (London: Diploma Thesis. Architectural Association Archive), p. 48.

and design methodology is discussed, but he is wary of taking a simplistic approach.

The actions that concern Evans can not be confined to discussing positive or negative interferences, however, for there are also internal resistances to the fulfillment or blockage of dreams and desires. There are many intervening variables, many imponderables and subjective implications impinging on actions that remain undiscovered by systems analyzers. Subsequently, Evans proclaims: goal-oriented human action can not be used as a design criterion. In other words, design is not a cybernetic feedback process. Instead, design is likened to an electrical circuit of flows and resistances. He argues that even though actions and intentions underlie everyday existence and are inextricably linked, they do not generate a design methodology. New intentions and actions tend to arise from novel situations producing unpredictable actions. Thus new physical or social systems produce unforeseeable human actions. Consequently, as time moves forward, there are always new additions to the set of possible actions and the bounds of freedom become larger and larger.

In addition, there are two time-based considerations that affect freedom of action. Compression of time leaves more free time for other activities. The road and rail systems, domestic equipment such as washing machines and vacuum cleaners, mail order catalogues, supermarkets, machine tools, and typewriters all compress the necessary time of traveling, cleaning, shopping, or communicating. On the other hand, time-consuming systems such as cameras, television, printed books, watches, money, second and third generation computers, and buildings also exist.

How, then, does Evans wrap his analysis around his interest in "goal achievement", and how does a given system frustrate or free individual actions? Daily life, he argues, is filled with interactions with "things" that are either permissive, allowing us to satisfy our needs, or restrictive, undercutting satisfaction. In other words, the physical world offers variable resistance to the accomplishment (i.e., the free flow of current) of desires, intentions, or goals. These variables of resistance make that achievement easy or difficult, and if too difficult there are surrogate goals to be achieved instead. Some of these resistances are low enough to be overcome; some desires may be too weak or the resistance too strong to enable action; and finally, only a substitute goal may facilitate action.

Action is how freedom is brought into a closer relationship with "things" for it is these "things" that make a society "permissive" or not. A designer does not satisfy needs by providing a "good" environment or providing a set of utilizable articles. Needs are far too ephemeral for that, and are conditioned by the resistances of custom, approbation, consent or habit. What the designer does is to reduce a series of resistances to desired action, which subsequently may be allowed or not. "Thus it is that the physical systems within and with which we exist are capable of affording greater or less control of our actions – greater or less volitional scope."[29] All of Evans' concerns are derived from problems just coming into awareness within emerging "Societies of Control" - problems over how a designer deals with communications that come up against resistances, that impede or facilitate freedom of choice and action.

There are externalities to be taken into account in any design process. The designer produces artifacts, tools and support-systems, and these systems tend to generate further processes. Evans makes a diagram showing how Siemens Bessemer steel making processes when first introduced into 19th century America generated many other changes. What irks Evans is that all of this analysis of changes and processes took place at a distance, without consideration for those whose lives were being changed. "Effective individual and social morphology does not occur in this way, but by subtle processes of infiltration and goal-reorientation...".[30] Information must flow both ways: from system designers, and from individuals who interact with these new systems. The inability to be interactive, or so Evans maintains, is the problem with the modern movement of architecture; it is attached to the logic of social manipulation; it intends to change people's patterns of action by designing physical systems to affect patterns of human development, but it does so at a distance, without considering human beings directly. Evans puts in an aside: fortunately due to the incompetence of planners such control interventions remain only intent not an actual effect. In any good systems analysis, there are a multitude of variables to consider in order to measure a system's effect on people's behavior and vice versa, not the simplistic one-way control mechanisms that planners invent.

Relying on the philosophical work of Henri Bergson, E. Schrodinger, and Norbert Weiner, Evans proclaims that man is the vanguard of "anti-entropy"; he is always desirous of greater order, imposing it on both inert physical systems and organic systems. Physical systems tend towards entropy, greater randomness and disorder, but when humans interact with physical systems, these systems tend over time to become more organized or anti-entropic; human intervention aims to achieve greater order not disorder. Evans believed this was wrong: "freedom" and "order" do not go hand in hand. Human patterns of action do not have to be designed to reduce the entropic processes of the entire physical system. Evans finally points to an example that apparently has motivated his elliptic discussion in the first place - a critique of planning published in *New Society* under the title of "Non-Plan: An Experiment in Freedom" authored by Reyner Banham, Paul Barker, Peter Hall and Cedric Price. He quotes from a letter to the editor criticizing the "Non-Plan" by comparing it to an experiment in child-raising. One group of

---

[29] Evans, *Towards Anarchitecture, op cit,* p. 27.
[30] *ibid*, p. 28.

children was fed a balanced diet at regular intervals (i.e., control theory); another group were allowed to roam about and feed themselves when they wanted (i.e., freedom enhancement). As far as the health and weight of the children were concerned, both groups achieved the same goal. But there was a big difference: the mess left behind by the random eaters required adults to clean it up.

Evans extended this discussion by noting that most designers opt for the first consideration and base their instinct to order the environment on conceptual simplicity. Evans prefers the messy choice on humanitarian grounds. "The utilitarian basis of architectural functionalism has tended to simplify notions of purpose, and has given us only the ankle-cartilage of what is a much more complex affair. It is surely time to effect a reconstitution of the corpus."[31] Architects sell-out if they opt to keep the game of design simple at the expense of coordinating with reality – they should not take averaged needs and desires as their guideline but consider the world for which they design as a stage for people's actions. In the end, the stage belongs to the people, and the designer should enhance their freedom and facilitate their actions. It is not a matter of removing all planning controls to enhance freedom as advocated by the "Non-Planners" and the "messy eaters", but designers might instead provide machines to clean up the mess and enhance freedom thereby.

Evans continued his interest in criticizing the architecture of control and focus on human behavioral, reporting in 1972 on a series of experiments undertaken by Dr. Roy Slim of the Department of Educational Behaviorism at the Maudlyn Institution entitled: E.A.T., or "Educational Aversion Therapy".[32] These experiments were in fact a shock therapy protocol designed for a group of London architecture and art students to test how to control their aesthetic taste. A series of images from antique furniture to 1960's mega-structures were shown to the students who were consequently administered tolerable levels of electric shock whenever an image appeared that they were supposed to consider a bad design. The negative association of an unpleasant shock with the image was intended to program their brains to make negative judgments and short circuit the normal association between cause and effect. Obviously an experiment in control theory, the good Dr. Slim acknowledged that "education is not just leaning about things. It is learning to ignore them....the whole world would cease to operate if everybody was open-minded. Nobody would know what to do anymore."[33]

What Evans objects to, is the deterministic and behvioristic nature of control and communication systems, eliminating alternatives, random acts,

---

[31] *ibid*, p. 32.
[32] Evans, Robin, "Punishing Architectural Prejudice, Directions in Design Education" *ARse* 5/6 (Spring/Summer 1972): 26 -7.
[33] Quoted by Irene Sue, quoting Robin Evans' "Punishing Prejudice", *ibid*, p. 27.

and individual resistances in order to achieve the one directed goal, and thereby avoid at all cost disorder and chaos. He substitutes, instead, freedom of choice and behavior, movement and change, pointing out the many resistances to any simplistic systems analysis of human behavior, drawing into consideration unaccounted variables that get in the way. Like Banham, Evans was interested in streams of information entering from other times and spaces, integrating these into a designers methodology.

Gordon Pask

Robin Evans' AA thesis project entitled "Toward Anarchitecture: Artifact Systems with Respect to Human Freedom", might also be seen in the light of the cyberneticians Gordon Pask's teachings at the AA in the 1960's, and his involvement with Cedric Price's and Joan Littlewood's conceptualization of "The Fun Palace". Price and Littlewood began their collaboration in 1961, involving the cybernetician Gordon Pask after 1964. They wanted to design an adaptive structure that would vary according to users' interests and programmatic considerations. The problem was how to insert improvisation into an architectural structure; how to design a structure that could be reprogrammed and reconfigured over time. In order to achieve such complexity, the "Fun Palace" project by 1966 became a huge collaborative effort involving programmers, cyberneticians, and structural engineers. Gordon Pask was head of the "Fun Palace Cybernetics committee", one of four consultant groups with the others focusing on structure, programming, and sociology.[34] Pask defined cybernetics as the study of how complex biological, social, or mechanical systems organize themselves, regulate themselves, reproduce themselves, and evolve over time and learn. He assumed there was constant communication or "conversations" between parts of any given system, between man and his environment. Committee meeting notes in Cedric Price's archive contain the following essay that borrows from Pask's learning theory:

> Man must learn. He enjoys living in a world that gives him enough to learn about without becoming utterly unintelligible in its variety. But his world is individually specified and because even the individual is in a flux of adaptation it is evanescent. So, for some purposes, the environment that a man enjoys must be automatically tailored to suit his changing attitudes just as, in a

[34] Mathews, J. Stanley, "An Architecture for the New Britain: The Social Vision of Cedric Price's Fun Palace and Potteries Thinkbelt" (Unpublished PhD dissertation, Columbia University, 2003) (UMI Number: 3074308).

> conversation, we continually modify the common language of our discourse.[35]

To erect such a self-adapting and user-responsive environment, the designers intended that computers be deployed throughout the "Fun Palace". These computers would gather and assess data from electronic sensors and response terminals; they would assign a prioritized value to the interests and activity preferences of individual users and use these parameters to modify spaces and activities within the structure; they would reallocate moveable walls and walkways to adapt the form and layout of the "Fun Palace" to changes in use over time. Expectations were that the building would learn behavioral patterns and be able to plan future activities according to a series of algorithms. In addition to computers, there were plans to install a number of cooperative machines operated by two people, individual teaching machines, closed-circuit TV's, and surveillance systems allowing participants to "experience the emotional thrill and power" of watching themselves participate.[36] A "Pillar of Information", designed for the Palace's main entry, was intended as an electronic kiosk that could search, display and track information of all sorts. The pillar would allow users to access other user's interests and queries, and would over time, develop a complex network of cognitive associations or develop into a nonhierarchical information matrix. Joan Littlewood believed "...[t]he Fun Palace is both a pleasure arcade and an instrument which motivates the typically passive participant into thinking more abstractly....The 'fun arcade' will be full of games and tests that psychologists and electronic engineers now devise for the service of war - knowledge will be piped through juke-boxes".[37]

Oblivious to the authoritative nature of control theories if pushed to extremes, Littlewood waxed poetically in a letter written to Gordon Pask in 1964:

> The operators in the social system are like mirth and sensuality. Its operators are actions or intentions or changes in the shade of joy or grief. We can to some extent control these transformations, though, in this case, we and our machinery act as catalysts and most of the computation is done as a result of the interaction taking place

---

[35] Price, Cedric, "Notes on Cybernetic Aspects" (1963) *Fun Palace document folio, DR 1995: 0188526 Cedric Price Archives, CCA Montreal.* Quoted by Stanley Mathews "The Fun Palace as Virtual Architecture: Cedric Price and the Practices of Indeterminacy", *Journal of Architectural Education* (2006): 39-48; quotation on p. 43.

[36] Price, Cedric, "Fun Palace", *op cit*, p. 45.

[37] Littlewood, Joan and Price, Cedric, "A Laboratory of Fun", *New Scientist* 38 (14 May 1964): 432-33; quotation on p. 432. Quoted by Lobsinger, Mary Louise, "Cybernetic Theory and the Architecture of Performance: Cedric Price's Fun Palace" in Williams Goldhagen, Sarah and Legault, Rejean, *Anxious Modernisms* (Cambridge: MIT Press, 2000), pp. 119-139; Quotation on p. 130.

> between members of the population, either by verbal discourse, or by competitive utilization of facilities, or by cooperation to achieve a common objective.[38]

Littlewood continues: "It is right, in a project of this kind, to advance beyond the bounds of respectability and to move into the hinterland of things…for we then will know a great deal about how to control people and how to make them happy."[39] In 1969, Gordon Pask, who had been teaching at the AA from time to time since 1965, published a short article in *Architectural Design* about "The architectural relevance of cybernetics."[40] This was a special issue on "thinking about architecture and planning" edited by Roy Landau, and contained companion articles by Cedric Price, Karl Popper, Chris Abel, Lancelot Law White, Warren Chalk and David Greene. They all displayed sensitivity to the fact that the computer, or cybernetics, had dramatically transformed architectural thought.

Pask began simplistically pointing out that PERT, CAD, and an early graphic design program were all evidence of "cybernetic" methods applied to architecture. But Pask believed the relationship went even deeper: cybernetics and architecture share a common philosophy in the sense that architects over the last 100 years have become system designers, they are now interested in non-tangible organizational properties of development, communication, and control.

Architecture exists to perform certain functions performed for human beings or human societies. A building, Pask continued, is meaningful only as a human environment, continuously interacting with its inhabitants, serving them or controlling their behavior. Pask called this "architectural mutualism", the mutual interaction between structures and men or societies. There are two consequences, as far as Pask was concerned, in this shift toward mutualism: an emphasis on form and the requirement to design dynamically – that is to consider that continuously over time a building regulates its human inhabitants.

Pask carried this systems analysis of architecture further. A structure is coupled with other dynamic systems – the city; for example, conceived in the context of its temporal extension (*e.g.*, its growth and development), interpreted in accordance with its intention, as a product of its plan. Furthermore, the structure of the city acts as a "symbolic control programme" that regulates its behavior and renders it homeostatic. He envisioned urban development as self-organizing systems in which architects predict the extent to which the growth of the city would be chaotic or ordered by differentiation.

---

[38] "Joan Littlewood Letter to Gordon Pask" *Price Archive*, box 1/5. Quoted by Lobsinger, *ibid*, p.133.

[39] Lobsinger, *op cit.* p. 133.

[40] Pask, Gordon, "The Architectural Relevance of Cybernetics", *AD* 39/9 (September 1969): 494-6.

Architects are asked to anticipate trends in the development of educational technology and provide for their eventual impact on whatever structure they are designing. Pask cites the "Fun Palace" as an example in the field of entertainment, but in general wherever the notions of control, communication and system enter the discussion, abstract cybernetics was being interpreted as architectural theory. As a result of information and control processes, architectural design took place at several interdependent stages. First, architects specified the purpose or goal of a system, always underspecified but nevertheless offering a set of constraints that facilitated desirable modes of evolution; second, architects selected environmental materials; third, they selected the set of invariants programmed into the system; fourth, they specified what the environment will learn and adapt to (*i.e.,* feedback control mechanisms); and fifth, they develop a plan for adaptation and development.

Again, Pask refers to the "Fun Palace" system as an adaptation project and he also mentions that an urban plan also is governed by reactive processes. In either case, it is the delineation and dynamic modification of a control program that forms the basis of Pask's cybernetic approach. But here, in a clever word play, Pask allows that the designer is controlling the construction of a control system, hence the result is the "control *of* control". The designer operates at a higher level of the organization – not as a controller *per se*, but as a mixture of catalyst, crutch, memory and arbiter.

Conclusion

The decades of the 1950's and 1960's witnessed the bare beginnings of "Societies of Control" that would eventually expand into a global logic. Historians, artists, and architects were infatuated with systems theory and the information processing that electronic computers promised; they eagerly tried to adapt these new theories to their design methodology, producing bare traces of ideas in gestation. Yet, from what layer of time and space are artifacts rescued, becoming the shape of time that historians record? How is perception transformed, negotiated, and contested; how does this interactivity, this time-oriented feedback loop, effect design methodology, expand or limit the field of choice, and control behavior? What is the relationship between software and hardware, the program and the machine, and how does this transform the process of design? How is information recorded, translated, and transmitted; its incompleteness and indeterminacy dealt with and how does this effect the writing of history and procedures of design? Will the historian and artist operate in a closed world of information exchange, or an open-ended indeterminate one? Communicative exchanges, the new information order of "Societies of Control" with their networks, series and nodes, added new layers to be discovered in the exploration of post-WWII chrono-topographies. Time, process, choice, command, control, communicate were concepts of engagement belonging to new modes of thought, transforming both architecture and the writing of history.

Digital Architecture and the Temporal Structure of the Internet Experience

Antoine Picon

The paper argues that the temporal structures involved in the digital experience constitute, in a similar way, unavoidable references for contemporary architectural designers. The Internet, in particular, is synonymous with a series of very specific temporalities the resonance of which may be traced in many architectural and urban proposals often characterized as digital. These temporal structures give precedence to events. Like the Internet which encapsulates some of its fundamental characteristics, this world seems to bathe in an ever-intensifying present deprived of a clear historical dimension. In this context, the crisis of traditional tectonics that can be observed through emblematic productions like the Sendai Mediatheque, the Yokohama Terminal, or the recently completed Stadium of Beijing, may have to do with a more fundamental difficulty of architecture to be related, as it used to be, to collective memory.

Introduction

Since the dawn of modernity, a large part of the temporal concerns expressed in architecture has had to do with the relation between the rhythms of production, use and possible obsolescence of buildings, and the tempo that was observed in technology. Breaking with the old ideal of permanence that had characterized the architectural discipline almost from its start, modern architecture was fascinated by the general speed of mechanized society. Architecture was no longer about building monuments ideally eternal like pyramids; rather, it was permeated by a desire to emulate mechanized speed by the use of techniques such as prefabrication and dry assemblage. The rapid shifts in customer tastes and the subsequent evolution of markets raised the issue of the temporary and ephemeral and their possible acclimatizing in architecture. Finally, the question arose of a possible embodiment of these various temporal structures in the vocabulary of architecture itself. In various occasions, the speed of cars and aircrafts found a translation in architectural forms and aesthetics. This phenomenon was, for instance, the case with the famous Obus Plan for Algiers by Le Corbusier with its aesthetic treatment of the motor freeway.

In this paper, I will argue that the temporal structures involved in the digital experience constitute, in a similar way, unavoidable references for contemporary architectural designers. The Internet, in particular, is synonymous with a series of very specific temporalities, the resonance of which may be traced in many architectural and urban proposals often characterized as digital.

To be clear, the model I am proposing is not based on the hypothesis that technology shapes architecture without counterpart. I am not techno-determinist insofar that I don't believe that the relation between technology and architecture is unidirectional, technology shaping architecture like an external factor, something like an infrastructure determining a superstructure. Technology is of course partly autonomous; but its autonomy is limited.

In the perspective adopted here, technology and architecture are rather rooted in a broader socio-cultural ground that determines them both, as well as it informs their mutual relations. Modern architecture was sensitive to the rhythms of industry because of this socio-cultural common ground. In the same way, the Internet and contemporary design bear the imprint of temporal categories that extend beyond them. In such a perspective, the Internet is a highly revealing field of observation.

In a similar way, although I will be dwelling heavily on digital design, I don't want to consider it as a kind of neo avant-garde, as something exceptional. I am far from the ambition to transform it into something completely specific. According to me, digital design appears rather as the expression of more general trends at work in the world of design and even beyond, since it shares its tools with other domains, from the sciences to the digital arts. In other words, just like technology, digital design is both determining and determined. It has a certain degree of autonomy, but again this autonomy is limited.

I should also specify that, although some of my developments may sound critical of digital design, I am actually far from the hostile positions taken by theorists like Kenneth Frampton or Alberto Perez-Gomez.[1] For me the advent of the digital is contemporary with a series of crisis of dimensions we used to believe as constitutive of architecture. But it does not necessarily mean that this crisis is inherent to the media. It has more to do with a period of profound transition that extends, once again, far beyond the use of the computer. In a way, my argument could be summarized as such: we are indeed in the middle of a dramatic redefinition of the relation of architecture to time and history. Yet, it is certainly not the end of this relation, even if the present situation presents something deeply unsettling.

A World of Events

It has become a common place to state that the generalization of the computer and more generally the development of digital culture are intimately linked to a dramatic reshaping of our perception of the world. In other words, digital

[1] See, for instance Frampton, Kenneth, *Studies in Tectonic Culture: Studies in the Poetic of Construction in Nineteenth and Twentieth Century Architecture* (Cambridge, Massachusetts: MIT Press, 1995) and Perez-Gomez, Alberto and Pelletier, Louise, *Architectural Representation and the Perspective Hinge* (Cambridge, Massachusetts: MIT Press, 1997).

culture is as much about experiencing and seeing, as it is about computing. But what does one see exactly, on a computer screen?

For a designer, the answer may seem simple at first: what one sees are forms. But actually, these forms are in constant flow until the moment the designer and his partners decide that they should become definitive. What one sees is rather a moment or a series of moments in a process organized along geometric flows. The pervasive presence of geometric flows explains the recurring reference made by such theorists as Greg Lynn to Muybridge, or Marrey in experiments with the recording of movement.[2]

In other words, what one sees on a computer screen is something that "happens". In the digital realm, form, architectural form, represents an occurrence; it happens. This manifestation is probably among the factors that explain the recent development of the performative approach in architecture.[3] For somebody working in the financial markets, this temporal, event-like structure of what one sees on a computer is even more evident. What a trader deals with using the latest digital equipment are situations on disputed markets that are comparable to battlefields.

From the start, digital culture was about seeing events. It is worth remembering that one of its first major applications, the North American antimissile system SAGE, designed under the direction of MIT computer scientist Jay Forrester, was meant to allow operators to see situations like a nuclear strike. The profound connivance between nascent digital culture and the Cold War had to do with the role they both gave to events and their possible integrations into scenarios. In the Cold War perspective as analyzed by historian Paul Edwards, the computer screen was an integral part of the war room.[4]

Nevertheless, the relation between digital culture and events runs even deeper. As the French philosopher Pierre Lévy remarked in a groundbreaking essay entitled *La Machine Univers*, a bit of information is not a "thing" but an "occurrence", an atomistic event.[5] This event corresponds to something that happens, rather than something that is following traditional ontological categories. For all that, this growing importance given to events did not belong to digital culture alone. It was actually part of a more general evolution, the early expressions of which can be traced also in the architectural and urban domain. In the 1950's and

---

[2] Lynn, Greg, *Animate Form* (New York: Princeton Architectural Press, 1998).

[3] Kolarevic, Branko and Malkawi, Ali M. (eds), *Peformative Architecture: Beyond Instrumentality* (New York: Spon Press, 2005).

[4] *cf.* Edwards, Paul, *The Closed World: Computers and the Politics of Discourse in Cold War America* (Cambridge, Massachusetts: MIT Press, 1996). The relation between the computer and the war room was treated in a spectacular way by director John Badham in his 1983 film *WarGames*.

[5] Lévy, Pierre, *La Machine Univers: Création, Cognition et Culture Informatique* (Paris: La Découverte, 1987), p. 124.

1960's, cities were for instance increasingly apprehended as systems of events. This system was, for instance, the leading theme of Archigram's 1963 *Living City* exhibition that opposed the vitality of the city as a structure enabling the production of occurrences, events, and scenarios to the sterility of the same city as envisaged by modern planners. Part of the exhibition message was to attract the attention of the public to the events, sometimes major, more often minor, that structured the city as a global environment.[6]

Archigram owed a lot to a sociologist like Henri Lefebvre, and his approach to the structures of everyday life, and above all to the Situationists' vision of the city as a living matrix of situations that could be explored using merely the drift and psychogeography.[7] The Paris evoked in the famous psychogeographic maps of Guy Debord and Asger Jorn was to be interpreted in that way. Furthermore, despite his exclusion from the Situationist movement, Constant was to remain faithful to this interpretation with his New Babylon project series that tried to give a spatial consistence to the notion of a city of events.[8] The megastructural movement in the 1950's and 1960's was also permeated by the same kind of intuition, for megastructures were often interpreted as devices meant to promote random encounters and situations. This approach was already the case with the Smithson's Golden Lane Project.[9] A designer like Yona Friedman had something similar in mind with his spatial cities in which people would wander, roam and meet like passengers on the deck of a liner.[10] The megastructural production of Archigram followed the same path. The Centre Pompidou, with its aerial promenade meant to foster encounters and conviviality, was probably one of the ultimate forms taken by this vision.[11]

Since that time, the spectacular development of digital media has reinforced our perception of cities not as much as traditional places, but as spaces or territories where things literally take place. Allied with surveillance technologies, digital media present us at various levels with the spectacle of what happens in our urban environment, from major events to minute occurrences like the notifications we get of traffic conditions on the road. Through the pervasive presence of digital media, our life is structured by thousands of events that organize our perception of cities, and more generally

[6] *cf. Archigram* (Paris: Editions du Centre Georges Pompidou, 1994).

[7] Sadler, Simon, *The Situationist City* (Cambridge, Massachusetts: MIT Press, 1998).

[8] Wigley, Mark, *Constant's New Babylon: The Hyper-Architecture of Desire* (Rotterdam: Witte de With, Center for Contemporary Art/010 Publishers, 1998).

[9] *cf.* Rouillard, Dominique, *Superarchitecture: Le Futur de l'Architecture 1950-1970* (Paris: Editions de La Villette, 2004).

[10] Lebesque, Sabine, *Yona Friedman: Structures Serving the Unpredictable* (Rotterdam: NAI Publishers, 1999),

[11] On the ambitions of Centre Pompidou, see Piano, Renzo and Rogers, Richard, *Du Plateau Beaubourg au Centre Pompidou*, a conversation with Antoine Picon (Paris, Editions du Centre Georges Pompidou, 1987).

the world, as "what happens". "What happens": this was the theme picked in 2002 by Paul Virilio for an exhibition and a book.[12] The main limitation of the project, however, was that it paid attention only to the most dramatic events of urban life, from a major flood to terrorist attacks, forgetting that our urban existence is structured by millions of more innocuous ones. These events include all the festivals, celebrations, and games that characterize the contemporary urban scene. They include also micro-occurrences like the weather forecast, which has been transformed over the past years into a kind of ritual, coloring daily life.

At this stage, I would like to introduce the following argument: too many events tend to neutralize the historical potential of events. In a world in which millions of things constantly happen under our eyes, it is actually as if nothing happens. It is now time to turn to the Internet as the best symbol of this paradoxical situation. For the Internet seems to bathe in an everlasting present made of a rich texture of situations, events, and scenarios. In other words, the Internet is a formidable archive, but an archive oblivious of its process of constitution and sedimentation.

Towards the End of History?

The Internet is the historical product of an evolution leading from realizations like the early-1970 Arpanet to the present cyberspace, an evolution in which the imagination of authors like William Gibson or Neil Stephenson played a role beside military concerns, corporate strategies, and alternative communities' experiments. In other words, some forty years after the early experiments that would lead to it, and after some fifteen to twenty years of semi-public and public use, the net has become a historical reality the evolution of which has begun to be studied.[13]

But despite this historical load, the net seems to thrive in the diffuse light of an everlasting present. On the Internet, pages appear and change without the visible sedimentation that characterizes changes and reallocation in the ordinary space. The Internet is an archive, but it does not sediment in space like traditional archives. Traces are erased everyday, and their memory seems for ever lost even if they remain somewhere, in the recesses of distant servers. The evolution of the network itself seems to belong to a different category of phenomena than the growth of traditional infrastructures. Now that it is covering most of the globe, it appears more akin to a never-ending intensification than to an extension.

In that respect, the Internet reproduces features characteristic of the world of globalization. Indeed, globalization is often perceived in the light of

[12] Virilio, Paul, *Ce qui Arrive* (Paris: Fondation Cartier pour l'art contemporain, Actes Sud, 2002).

[13] See for instance Abbate, Janet, *Inventing the Internet* (Cambridge, Massachusetts: MIT Press, 1999); Flichy, Patrice, *The Internet Imaginaire* (Cambridge, Massachusetts: MIT Press, 2007).

a never-ending present, like a process of indefinite intensification. In his bestseller of the early 1990's, *The End of History and the Last Man*, Francis Fukuyama offered one of the most striking expressions of this approach.[14] Despite the multiplication of conflicts of all kinds, the lack of a clear alternative to the capitalistic world as it is, leaves us prone to the same kind of impression.

This world, our world, is indeed full of dynamism. Political regimes come and go, wars erupt and turn bad, new sources of profit emerge and corporations are created while entire sectors of the economy decline and disappear. Somewhat similarly, on the Internet things constantly happen. The Internet is the ultimate expression of an electronic world which is in its very substance nothing else than billions of occurrences, bits of information that are created and circulated. But the multiplication of things that happen creates a scintillating giant texture rather than a clear series of patterns. It is, by the way, striking that one of the most common visual metaphors associated with the Internet has to do with a city, or a strip seen at night from above. The Internet as a giant Las Vegas, centered on its animation and somewhat oblivious of the possibility of an historical evolution: the metaphor was already present in Gibson's *Neuromancer* or in Stephenson's *Snow Crash.*[15] This metaphor is ever more pervasive today.

While the Las Vegas strip is one of the analogies that have been mobilized regularly to give an intuitive characterization of the Internet, especially at its beginning, many theorists have put together the development of digital culture and the radical turn taken by suburbanization in the developed countries. The digital condition announced by Nicholas Negroponte has indeed to do with dispersion, a dispersion counterbalanced by the instant mode of communication made possible by the Internet.[16] The theme of dispersion is implicitly present in the choices made for the covers of a great deal of books dealing with the spatial impact of digital culture, from Joel Kotkin's *New Geography* to William Mitchell *E-Topia.*[17] It is as if the world has become a giant garden where the population is loosely distributed. Some of the key places where digital culture is being produced are indeed suburban, such as Silicon Valley or Cambridge, Massachusetts. The "eternal present" of the web and its suburban flavor recall the world evoked in the American movie *Pleasantville*, a 1950's TV-show-universe in which things constantly occur, episode after episode, while nothing really important ever

[14] Fukuyama, Francis, *The End of History and the Last Man* (New York: Free Press, 1992).
[15] Gibson, William, *Neuromancer* (New York: Ace Book, 1984); Stephenson, Neil, *Snow Crash* (New York: 1992, new edition New York: Bantam, 2003).
[16] Negroponte, Nicholas, *Being Digital* (New York: A. A. Knopf, 1995).
[17] Kotkin, Joel, *The New Geography: How the Digital Revolution is Reshaping the American Landscape* (New York, 2000, new edition New York: Random House, 2001); Mitchell, William, *E-Topia: 'Urban Life, Jim — But not as we Know It'* (Cambridge, Massachusetts, 1999, new edition Cambridge, Massachusetts: MIT Press, 2000).

happens.[18] *Desperate Housewives* could be a contemporary equivalent of that world. With the Internet, saturated with information on what happens all over the planet, we seem to be living in some kind of suburban suspension of history.

The only possible return to history that seems conceivable in such a context is actually the real end of history, a global catastrophe that would terminate the present as well as eradicate the remnants of the past and the promise of a future. This structure was already in place at the beginning of what was to become the Internet. For, as I already stressed, early computer networks like the SAGE system were elements of a strategy of calculated immobility meant to avoid a nuclear holocaust. Throughout the 1970's, the Cold War perspective of nuclear war was complemented by the dismal perspective of the exhaustion of natural resources, beginning with oil. Revealingly, Jay Forrester, the creator of the SAGE system, was among the early explorers of this exhaustion of natural resources. His world dynamic computer simulation served as the basis of the pessimistic scenarios popularized at the time by the Club of Rome.[19]

More recently, global warming has replaced these early forms of Armageddon, but the perspective remains fundamentally the same: a game-like present reality, based on the intensification or saturation of what is, in the shadow of an ominous catastrophe. Awaiting the final catastrophe in an everlasting present populated with events of all kinds, we live in a world that is not only reminiscent of TV shows. It is also strangely akin to the heroic fantasy universe. Our world is indeed magic, full of surprises, changes of direction and metamorphoses. Just like heroic fantasy fiction is always placed under the aegis of an ominous threat; we are living, so to say, under the volcano. It is no coincidence that from cinema to video games, from the *Lord of the Rings* to *Harry Potter*, digital culture has given new credentials to the heroic fantasy genre.

Designing for the Present

One is struck by the way entire regions of contemporary architecture have inserted themselves in the temporal structure I just evoked. Never has architecture been as obsessed by dynamism, movement, flows, as today. As I mentioned earlier, never have the dynamic images produced by photographers and scientists such as Muybridge or Marrey been so present in the architectural imagery. Yet architecture paradoxically appears at the same time strangely disconnected from any real kind of historical perspective. The best expression of it is perhaps the importance taken by the notion of field. "The flow of historical conditions" that many architects try to deal with is

---

[18] Ross, Gary (director), *Pleasantville*, 1998.
[19] *cf.* Edwards, *op cit.*

actually interpreted as a surface, [20] a field of material forces in which the gradient has rather to do with capital and space than with time proper, or rather historical time. In the theoretical debate, the notion of "post-criticality" advocated by authors like Robert Somol and Sarah Whiting, is related to this suspension of historical time.[21]

The use of the computer has allowed these notions of field, flow, and gradient to be translated into architectural forms. The irony is that the temporal processes these forms are supposed to refer to has been replaced by geometric manipulations. Contemporary digital architecture is full of frozen flows, waves, and other turbulences. Intensity, an intensity of purpose and sensation-inducing effects, is replacing meanwhile the perspective of historical evolution. This impression of and everlasting architectural present is of course related to the disconcerting newness of some of the forms produced with the computer. Even if we are now past the early blobs, many digitally-produced forms do not clearly belong to well-identified evolutionary processes starting from clearly identified precedents, as variations on well-rehearsed codes. Whereas mannerism and baroque were evidently re-elaborations of the Renaissance fundamentals, today's architectural geometries have no clear precedent from which they derive, despite the attempts to relate them to Guarini, Borromini and other forerunners of architectural complexity. In that respect, modernism, despite its claim to the contrary, still emerged and developed in a frame based on codes, precedents, evolutions, and variations. We are far today from being able to identify those elements in the digital domain.

What is also suspended in such a context is the traditional modernist obsession with the construction process as a possible source of expressive content. The dynamic forms of today's architecture have not much to do with the construction process itself. They are often even foreign or adverse to it, despite the fact that aided prototyping and fabrication are among the perspectives made possible by the use of the computer. Of course, such a situation might very well be temporary. New codes are probably emerging. But their lack of visibility, coupled with the absence of clear historical perspective, is deeply unsettling.

Real Time and Deferred Communication

A few years ago, the rise of online life was still presented as the triumph of real time over deferred communication. The latest in a series of technological innovations that had begun with the telegraph, the Internet was supposed to

---

[20] *cf.* Sanford Kwinter's contribution to *Flying the Bullet, or when did the Future Begin?* (New York, Princeton Architectural Press, 1996).

[21] Somol, Robert and Whiting, Sarah "Notes around the Doppler Effect and other Moods of Modernism", *Perspecta*, no.33, (2002): 72-77.

abolish waiting times and introduce everybody to the experience of being directly in contact with the pulse of the entire planet.[22]

We now know that things are a little more complex, that the Internet is developing according to the articulation of two very different kind of temporal regimes. Surfing, shopping, chatting, and playing online: these activities take place in an almost real time. Yet a lot of other activities are actually less immediate. Email is, for example, really useful because of the disconnection between sending and receiving times that gives a supplementary degree of freedom to its users. Deferred communication also rules activities like agenda coordination between firm-owned computers and personal digital assistants. Deferred time applies above all to all those processes that allow information to accumulate, waiting for the proper time and user. Despite the multiple erasures that characterize its day-to-day life, the Internet is both a place of communication and an archive, a strange archive for sure because of its glib, to say the least, relation to memory keeping. This dual model of the immediate impact and the deferred communication haunts also architecture and urban design. The desire for real time can take naïve forms, like the massive use of screens to foster cyberspace-oriented experiences. It expresses itself more often through the research of an almost tactile presence of form and materials. Sensation seems to be the 3D equivalent of online real time communication. The return of the ornament, advocated by design practices as different as those of Greg Lynn and Foreign Office, might very well have to do with the quest for a new architectural immediacy.[23]

This immediacy is however questioned as soon as it appears excessive. For architecture also points towards the deferred and the coded that needs patient deciphering. This might be one of the reasons so many architects criticize Frank Gehry for a seduction without reserve. In the 3D world, the deferred has to do with the need to reinstall the architectural object into an interpretative frame that is not immediately accessible through sensory experience. Another way to put it is to understand the "deferred" as the kind of opacity that reintroduces something like a resistance of architecture to its public. Deferred reception might represent the new inertia replacing gravity as the fundamental dimension of the architectural object. To be more accurate, many architectural proposals today deal with the intertwining of the immediate and the deferred, the sensual and the more cognitive-oriented, just like prior designs often dealt with the interlacing of the aerial and the heavy.

---

[22] For a critique of this mythical dimension of online life, see Mosco, Vincent, *The Digital Sublime: Myth, Power and Cyberspace* (Cambridge, Massachusetts: MIT Press, 2004).

[23] See for instance Moussavi, Farshid and Kubo, Michael, *The Function of Ornament* (Barcelona: Actar, 2006).

This duality has of course to do with the fundamental ambivalence of the subject of the digital experience. On the one hand, this subject likes to define itself through the irreducibility of sensorial experience to any program. On the other hand, the capacity of this same subject to decipher complex codes seems as important to his self-definition. Such a duality carries with it a specific form of anxiety bordering on paranoia. Am I a "real" subject or an intelligent piece of program, hence my capacity to decipher; am I autonomous, or even free, or totally determined by administrative and commercial apparatuses that extend far beyond my limited universe? This question, that is at the core of so many contemporary Hollywood productions, like *The Matrix* series, now looms in the architectural realm. Another way to characterize the phenomena I have been evoking so far, is to say that it is as if the inscription of architecture in a clear historical framework is often replaced by a permanent oscillation between the temporal regimes of immediacy and deferred communication. A series of rapid shifts invested with an almost hypnotic power because of their high frequency and intensity tends to substitute itself to a more traditional reference to historical categories. Even "the New", once worshiped by the Moderns, is no longer a valid reference. For sure, many computer-produced architectural forms look startlingly new. Yet this newness is far less relevant than the realm of sensations they are supposed to introduce.

The constant oscillation between a regime of immediacy and a regime of deferred communication also complicates the relation between the architectural object and the building process. The relation between architecture and its realization can no longer be apprehended in terms of a gradual unfolding of ideas and means. It is a complex mix of radical short circuits between design and fabrication and distances between the realms of conception and realization.

The Crisis of Memory

The Internet is an archive, but it is a strange archive in which what is kept is not necessarily what used to be considered irreplaceable in former depositories. During its evolution, from the initial Arpanet network to its present structure, the Internet has changed its "backbone" several times, like a giant library able to reorganize periodically the general structure of its collections without any visible impact on its readers.[24] One may also think of some strange form of sloughing in which the skeleton rather than the skin is abandoned. Contrary to former networks, the most fundamental level of the Internet is not constituted by a core of servers linked one to another but rather by the multitude of connected users. This feature is related to a more general

[24] *cf.* Abbate, Janet, *op cit.*

crisis in the traditional notion of structure, as well as to a simultaneous evolution of what we mean by infrastructure.

Regarding structure, it is striking to observe how this notion is less and less pertinent today. One of the reasons has to do with the fact that a lot of problems that used to be treated at the level of structural design are now transferred to the design of materials. This fact is evident with the car bumper, which used to be a structure based on a material that absorbs part of the energy of deformation. What we are observing today, in contrast, is the end in many cases of the rationalist quest for structure, the end of the gothic cathedral ideal that had been placed from Viollet-le-Duc on at the core of the modernist attitude towards edification.[25] Striking, the new technological universe produced by this weakening of the traditional structural evidence is full of soft surfaces. When these are not surfaces, they are rather more like webs than traditional structures. The Beijing Stadium of Herzog and De Meuron is typical of this evolution, which has found its theorist in engineer Cecil Balmond.[26]

Regarding infrastructure, in the present digital age, in many cases what is infrastructural is not the "backbone" or hardware, but rather the soft layers of programming and individual connections. In many firms today, the management software and the mailing lists are more essential than the plants and other physical infrastructures. In everyday life, changing the operating system, passing for instance from Macintosh to Windows, makes for a more fundamental decision than changing one's computer. The Internet is the best illustrations of this inversion. What is truly infrastructural is not the "backbone" at all. This is where Al Gore went wrong with the metaphor of the freeways of information and its claim that his vision shaped the development of the Internet.[27] It is, once more, the dense texture of users and the logic governing their exchanges that has really made the Internet what it is. This context might very well represent one of the most formidable challenges for architecture insofar as it tends to desegregate its traditional hierarchies. In other words, we are confronted with a deep crisis of tectonics. Tectonics was where the structural met with the legible and the symbolic. The crisis of traditional structural assumption implies a crisis of tectonics, perhaps even its end, if we are to follow Frampton's pessimistic assessment of the impact of digital culture on architecture.

One may of course wonder whether this is really the end of tectonics. After all, Beijing Stadium might indicate the path towards a new tectonics. Another way to put it then, would be to evoke the end of a certain type of

---

[25] At least, I might add, if we are to follow Kenneth Frampton on that point. See Kenneth Frampton, *op cit.*

[26] Balmond, Cecil, *Informal* (Prestel: Munich, 2002).

[27] Gore, Albert, "Remarks on the NRE", *Educom Review*, (Summer 1990): 12-16.

tectonics and its replacement by another one.[28] It is however striking how a certain indifference towards structure has developed in the past decades. Frank Gehry is perhaps the best example of this. His glib attitude towards structure is quite typical of the period. Indeed, traces of a Gehry-like glib attitude can be seen pretty much everywhere. The phenomenon is, in my opinion, rooted in the growth of cultural uncertainty, at a much broader level, concerning what sense to give to the structural and even the infrastructural. What do these terms mean today? In many cases the answer is far from clear.

Is it for that reason that so many iconic buildings are repudiating the traditional association between architectural expression and tectonics? Toyo Ito's Mediatheque was supposed to evoke an aquarium in which weeds floated.[29] The realized building has retained part of the initial ambition, but it is actually made of heavy duty steel plates that are more akin to ship construction. A similar distance between the soft fabric suggested by the initial presentation and the reality of the building can be observed in the case of the Yokohama Terminal. From Toyo Ito's Sendai Mediatheque to Foreign Office Architects' Yokohama Terminal, there seems to be no alternative but to radically distinguish the spheres of architectural expression, construction and tectonics. There is something quite paradoxical in observing, how, on the one hand, the computer allows one a better articulation of conception and construction, but on the other hand, creates a striking distance between the architectural imagery and the reality of the building techniques. More generally, the next step after deconstruction seems to be a state of suspension from traditional tectonic assumptions.

Among the consequences of the suspension of the traditional tectonic assumptions, one finds, as noted earlier, the increasingly determining role played by the ornament, a new ornament that is often comparable to a surface or field condition. One finds also a new interest taken in materiality.[30] Seemingly, there is actually a strong link between the development of digital culture and the widespread interest in materiality. The recent work of Herzog and De Meuron is quite emblematic of that connection. From the Basel Schaulager to the San Francisco De Young Museum, they have made an intensive use of pixellization, a technique directly linked to the use of the computer.[31]

---

[28] Beside Balmond's book, see for instance Leach, Neil; Turnbull, David; Williams, Chris (eds.), *Digital Tectonics* (London: Wiley-Academy, 2004); Reiser, Jesse and Umemoto, Nakano, *Atlas of Novel Tectonics* (New York: Princeton Architectural Press, 2006).

[29] *cf.* Witte, Ron (ed.), *Toyo Ito: Mediatheque of Sendai* (Munich: Prestel, 2002).

[30] This new interest in materiality is for instance well-conveyed by Toshiko Mori. On its link with the digital dimension, see Picon, Antoine, "Architecture and the Virtual: Towards a New Materiality?", *Praxis: Journal of Writing+Building*, no 6, (2004): 114-121.

[31] For a penetrating study of that question, see Rouyer, Rémi, "*Architecture et Procès Technique: Les Figures de l'Imaginaire*", typewritten PhD dissertation (Paris: Université de Paris I, 2006).

Yet the question remains as to the link between these new dimensions and memory. Such a connection cannot be constructed, in my opinion, but we are still at the beginning of this process. This link was, on the other hand, always evident in the case of traditional tectonics. Tectonics had to do with questions like the origin and the development of the arts, as had been made clear by the Abbé Laugier in the mid-eighteenth century.[32] The question remains fundamental, in the case of dealing, for instance, with the interpretation of the Greek Doric and its alleged filiation with wood construction, upon which historically theorists sharply disagreed.

Tectonics had a link with these questions through the question of the ruin, too. The ruin raised in a direct an unambiguous manner the question of the relation between architecture and time. What the ruination process revealed ultimately was the tectonic dimension of buildings. Above all, through its articulations that cannot properly be constituted as a language, yet are of a metalinguistic nature, tectonics related to time, history, and memory. Tectonics, in fact, held a protracted discourse on the very possibility to construct an argument about time, history and memory. A relationship that Erwin Panofsky had figured out in his famous essay on the relation between gothic architecture and scholastic thought.[33]

Furthermore, from the nineteenth century onwards, the link between architecture and memory was often doubled by a connection to the privileged media of memory and history. Despite Victor Hugo's famous statement that "*ceci tuera cela*",[34] that writing and printing had replaced architecture as the privileged instrument of collective memory, nineteenth-century architecture was actually trying to articulate itself with respect to writing. Labrouste's Bibliothèque Sainte-Geneviève is typical of that endeavor with its endless lists of famous people inscribed on its external walls.[35]

In the end, can digital media contribute to a permanent inscription? This might constitute an interesting question to raise today, especially since the digital is more akin to "events" than to "traces". Another approach to the question of memory may have to do with other dimensions to invent or reinvent in the digital context. What could be, for instance, the equivalent of weight and inertia in the digital world? The question of memory is actually crucial if we want to reinvent the perspective of a future different from the

---

[32] See Herrmann, Wolfgang, *Laugier and Eighteenth Century French Theory* (London: Zwemmer, 1962).

[33] Panofsky, Erwin, *Gothic Architecture and Scholasticism* (Latrobe: The Archabbey Press, 1951).

[34] Hugo, Victor *Notre-Dame de Paris* (Paris, 1831, new edition Paris: Le Livre de Poche, 1998), p. 289.

[35] On the relation between nineteenth-century architecture and writing, see for instance, Bergdoll, Barry, *Léon Vaudoyer: Historicism in the Age of Industry* (New York: Architectural History Foundation, Cambridge: Massachusetts, M.I.T. Press, 1994).

past. As historians like to point out, without a remembrance of the past, we are indeed trapped into the present.

Can we really live in an eternal present? This query is perhaps the real question on which I would like to conclude. The answer is probably negative. One shouldn't forget, meanwhile, that the eternal present used to be among the key characteristics of utopia. Utopia, indeed, announced the end of history, and its replacement by an everlasting present of felicity. Interestingly, utopia, or at least the demiurgic utopias that seduced modernism, has presently been discarded by architecture. In the past decades, following the injunctions to be "realistic" by Rem Koolhaas and others, architecture has been obsessed with joining the forces that truly transform reality instead of dreaming of alternative perfect cities. Time has become, by the same token, a major concern of contemporary architecture. Yet this time is at the same time, frozen.

Today's digital architecture is both obsessed by time, and strangely functioning as if its' flow had no serious consequence - that is until the ultimate ecological D-Day. The French philosopher Alain Badiou characterizes this situation as the convergence between a conservative agenda and an obscure longing for catastrophe.[36] There is perhaps something deeply true in this diagnosis. We are indeed petrified at the very thought that things could really change, and are in anxious anticipation of Armageddon.

Urgently, the necessity arises to deconstruct this complex Gordian knot, for architecture is not *only* about the present and the stabilization of existing modes of uses. Rather, architecture is as much about the possibility of a different future. Perhaps the time has come to be interested again in this different, alternative future.

---

[36] Badiou, Alain, *l'Ethique: Essai sur la Conscience du Mal* (Paris: Hatier, 1993), p. 36.

List of Contributors

Richard Arthur is a Professor of Philosophy at McMaster University, Hamilton, Ontario, Canada. Arthur specializes in the history and philosophy of science and mathematics, and specifically issues dealing with time, space, mathematics, and early modern philosophy. He has published an edition of translations of Leibniz's writings entitled: *The Labyrinth of the Continuum: Writings of 1672 to 1686* (New Haven: Yale UP, 2001), and numerous articles and book chapters on Descartes, Leibniz and Newton. He has also written on issues in the philosophy of modern physics, his most recent being: "Minkowski Space-time and the Dimensions of the Present" in Dieks, Dennis (ed.); *The Ontology of Spacetime* (Amsterdam: Elsevier, 2006), pp.129-155.

M. Christine Boyer is the William R Kenan Jr Professor of Architecture and Urbanism, Princeton University School of Architecture. Christine Boyer's most influential book to date is perhaps *CyberCities: Visual Perception in the Age of Electronic Communication* (Princeton, N.J.: Princeton Architectural Press, 1996) wherein she signaled, far in advance of most other scholars, the influence of communication technologies upon the global city. In addition, she has written many articles and lectured widely on the topic of urbanism in the 19th and 20th centuries. She is currently writing a book on Le Corbusier's writings entitled *Le Corbusier: Homme de Lettre*, and a series of collected essays entitled: *Twice-Told Stories: City and Cinema*; as well as a series of essays on contemporary urbanism entitled *Back to the Future: The City of Tomorrow*.

Nader El-Bizri is an Affiliated Lecturer at the University of Cambridge, Department of the History and Philosophy of Science, and Research Associate in Philosophy at the Institute of Ismaili Studies, London, as well as Chercheur Associé, Centre National de la Recherche Scientifique, Paris. El-Bizri is co-editor of the Kluwer Academic Publishers (Springer) book series on *Islamic Philosophy and Occidental Phenomenology*. His research interests include Arabic sciences and philosophy, phenomenology, philosophical and architectural theories of space and perception. His most important book length publication to date is *The Phenomenological Quest Between Avicenna and Heidegger* (Binghamton,: SUNY UP, 2000), as well as numerous papers on the phenomenology of space and time, including: "Qui-êtes vous Khôra?: Receiving Plato's Timaeus", *Existentia Meletai-Sophias* 11 (2001): 473-490; "A Phenomenological Account of the Ontological Problem of Space" in *Existentia Meletai-Sophias* 12 (2002): 345-364; "Ontopoièsis and the Interpretation of Plato's Khôra" in *Analecta Husserliana* 83 (2004): 25-45.

Mary Lynne Ellis is an analyst in private practice in London. She has Master degrees in Art Psychotherapy as well as Modern European Philosophy. She has lectured and published widely on phenomenological perspectives on psychoanalysis. Ellis is also a practicing artist. In her recent book, *Time In Practice* (London: Karnac, 2008), Ellis critically discusses implicit and explicit psychoanalytic concepts of time particularly in the work of Freud, Jung, Klein, Lacan, and Winnicott. She argues for the crucial relevance of theories of temporality of modern European philosophers such as Merleau-Ponty, Levinas, Foucault, and Ricoeur and also of the autobiographical writing of Eva Hoffman to contemporary psychoanalytic practices. Her most recent book is with Noreen O'Conner, *Questioning Identities: Philosophy in Psychoanalytic Practice* (London: Karnac, 2010).

Josef Früchtl is Professor of Philosophy of Art and Culture at the University of Amsterdam and the Chair of the Institute of Philosophy. Früchtl is one of the editors of "*Zeitschrift für Ästhetik und Allgemeine Kunstwissenschaft*" (Meiner Verlag, Hamburg). His research centers on the intersection of aesthetics, ethics, theory of modernity, and recently film theory, with three of his most important publications being: *Ästhetische Erfahrung und moralisches Urteil. Eine Rehabilitierung* (Frankfurt am Main: Suhrkamp, 1996), *Mimesis - Konstellation eines Zentralbegriffs bei Adorno* (Würzburg: Königshausen & Neumann, 1986), *and Das unverschämte Ich. Eine Heldengeschichte der Moderne* (Frankfurt am Main: Suhrkamp 2004). The English translation, *The Impertinent Self* has now also appeared (Stanford, CA: Stanford UP, 2009).

Peter Galison is the Joseph Pellegrino University Professor, and Director of the Collection of Historical Scientific Instruments at Harvard University, Department of the History and Philosophy of Science. Galison's main work explores the complex interaction between the three principal subcultures of twentieth century physics -- experimentation, instrumentation, and theory. Recent publications include *How Experiments End* (1987), *Big Science* (1992), *The Disunity of Science* (1996), *Image and Logic: A Material Culture of Microphysics* (1997), *Picturing Science, Producing Art* (1998), *The Architecture of Science* (1999), *Atmospheric Flight in the 20th Century* (2000), *Scientific Authorship* (2003), and most importantly, *Einstein's Clocks, Poincaré's Maps* (2003), a study of the ways in which the theory of relativity stood at the crossroads of technology, philosophy, and physics. Professor Galison's most recent publication is *Objectivity* (Boston: Zone Books, 2007).

Raviv Ganchrow is both an architect and a sonologist based in Amsterdam. Raviv Ganchrow's work focuses on interrelations between sound and space, aspects of which are explored through sound installations, writing and the development of sound forming technologies such as Wave Field Synthesis. He is currently teaching architectural design at the Faculty of Architecture TUDelft, and is a faculty member at the Institute of Sonology at the Royal Conservatory of Music in the Hague, the Netherlands.

Leslie Kavanaugh is both an architect and a philosopher, receiving her doctorate in metaphysics from the University of Amsterdam. For over twelve years she was a Senior Researcher (Reader) specializing in the philosophy of space and time at TUDelft, the Netherlands. Kavanaugh published: *The Architectonic of Philosophy: Plato, Aristotle, Leibniz* (Amsterdam: University of Amsterdam Press, 2007). Other publications include the edited volume with Arie Graafland; *Crossover* (Rotterdam: 010 Publishers, 2006), and most recently *Meditations on Space* (Amsterdam: 2010). Kavanaugh is the founder and director of studiokav.com in Amsterdam, the Netherlands.

Chung-lin Kwa is an Associate Professor in the History and Sociology of Science at the University of Amsterdam. After studying biology, Kwa completed his dissertation on a study of the development of systems-ecology in the context of twentieth-century engineering rationality. In 1999, Chunglin Kwa was a Global Environmental Assessment Faculty Fellow at the Belfer Center for Science & International Affairs at Harvard University. Kwa teaches courses in history of science, and chemistry and society. Shortly, he will publish in English translation, *The Discovery of Knowledge: the Other History of Sciences.*

Patricia Locke holds the Adolph Schmidt Chair, and teaches philosophy and literature at St. John's College, Annapolis MD. She has published her poetry as well as articles on the work of Merleau-Ponty, Hegel and Proust. She currently is completing a book length manuscript, *Recollecting Architecture: A Phenomenology of Ambiguity.* Her philosophical orientation comes from both the phenomenological tradition and an extensive study of French literature and ancient Greek culture. She has recently published, "Among the Hawthorns: Marcel Proust and Maurice Merleau-Ponty," in *Intertwinings: Interdisciplinary Encounters with Merleau-Ponty*, ed. Gail Weiss (Albany: SUNY Press, 2008) and is collaborating on a collection of essays: *Merleau-Ponty and the Experience and Expression of Space,* with Rachel McCann.

Sander van Maas is Professor of Musicology at Utrecht University and Assistant Professor of Musicology at the University of Amsterdam in the Netherlands. He received his doctorate with a dissertation on Olivier Messiaen and post-secular philosophy from the University of Amsterdam. He is Chairman of the Dutch Association of Aesthetics. His publications include, *The Reinvention of Religious Music: Olivier Messiaen's Breakthrough*

Toward the Beyond (New York: Fordham University Press, 2009), and *Faith in Music: On Olivier Messiaen* (New York: Fordham University Press, 2008). His current research project is entitled*: The Sense of Soul: On Being Music.*

Antoine Picon is Professor of the History of Architectur and Technology, and Director of Doctoral Programs at the GSD, Harvard University. Picon received engineering degrees from the *Ecole Polytechnique* and from the *Ecole Nationale des Ponts et Chaussees*, an architecture degree from the *Ecole d'Architecture de Paris-Villemin*, and a doctorate in history from the *Ecole des Hautes Etudes en Sciences Sociales*. Picon is best known for his work in the history of architectural technologies from the eighteenth century to the present, embedding the history of architecture within the social construct of knowledge. His *French Architects and Engineers in the Age of Enlightenment* (1988; English translation, 1992) is among his best known works in English, but he has published a series of other books: *Claude Perrault (1613-1688)*; *Ou la curiosité d'un classique* (1988), *L'Invention de L'ingénieur moderne, L'Ecole des Ponts et Chaussées 1747-1851* (1992), *La ville territoire des cyborgs* (1998), and *Les Saint-Simoniens: Raison, Imaginaire, et Utopie* (2002), and *Marc Mimram Architect-Engineer* (2007). In the past years, he has written various articles on digital culture and architecture, and he is currently preparing a book on the subject.